Student Solutions Manu
for
Thornton & Rex's

Modern Physics

FOURTH EDITION

Prepared by

Allen P. Flora
Hood College

Andrew Rex
University of Puget Sound

Australia • Brazil • Japan • Korea • Mexico • Singapore • Spain • United Kingdom • United States

© 2013 Brooks/Cole, Cengage Learning

ALL RIGHTS RESERVED. No part of this work covered by the copyright herein may be reproduced, transmitted, stored, or used in any form or by any means graphic, electronic, or mechanical, including but not limited to photocopying, recording, scanning, digitizing, taping, Web distribution, information networks, or information storage and retrieval systems, except as permitted under Section 107 or 108 of the 1976 United States Copyright Act, without the prior written permission of the publisher.

For product information and technology assistance, contact us at **Cengage Learning Customer & Sales Support, 1-800-354-9706**

For permission to use material from this text or product, submit all requests online at **www.cengage.com/permissions**
Further permissions questions can be emailed to **permissionrequest@cengage.com**

ISBN-13: 978-1-133-11219-8
ISBN-10: 1-133-11219-6

Brooks/Cole
20 Channel Center Street
Boston, MA 02210
USA

Cengage Learning is a leading provider of customized learning solutions with office locations around the globe, including Singapore, the United Kingdom, Australia, Mexico, Brazil, and Japan. Locate your local office at: **www.cengage.com/global**

Cengage Learning products are represented in Canada by Nelson Education, Ltd.

To learn more about Brooks/Cole, visit **www.cengage.com/brookscole**

Purchase any of our products at your local college store or at our preferred online store **www.cengagebrain.com**

Printed in the United States of America
1 2 3 4 5 6 7 16 15 14 13 12

Table of Contents

© 2013 Cengage Learning. All Rights Reserved. May not be scanned, copied or duplicated, or posted to a publicly accessible website, in whole or in part.

Preface and Acknowledgments

This book contains solutions to selected problems presented in *Modern Physics for Scientists and Engineers, Fourth Edition* by Thornton and Rex. Instructors who want students to have this manual, either as a required or optional text, may order it for their college or university bookstore through Cengage Publishing.

To the students using this manual, we want to remind you that there are many solutions to each particular problem. The one that you create is probably the one that you will understand best. However, the solutions presented here should provide enough detail to help clarify any problem with which you may have difficulty.

We would like to thank Stephen T. Thornton who offered suggestions for solutions to many of the problems. We also especially thank Paul Weber of the University of Puget Sound and Thushara Parera of Illinois Wesleyan University who checked these solutions and offered many corrections and suggestions for clarification.

For most physics problems, alternate solutions are possible. We welcome suggestions of alternate solutions and especially identification of any errors in this text, which are ours alone.

Allen P. Flora
Department of Chemistry and Physics
401 Rosemont Avenue
Hood College
Frederick, MD 21701
flora@hood.edu

Andrew Rex
Physics Department CMB 1031
University of Puget Sound
Tacoma, WA 98416-1031
rex@pugetsound.edu

© 2013 Cengage Learning. All Rights Reserved. May not be scanned, copied or duplicated, or posted to a publicly accessible website, in whole or in part.

Chapter 2

6. Let n = the number of fringes shifted; then $n = \frac{\Delta d}{\lambda}$. Because $\Delta d = c(\Delta t' - \Delta t)$, we have

$$n = \frac{c(\Delta t' - \Delta t)}{\lambda} = \frac{v^2(\ell_1 + \ell_2)}{c^2 \lambda}.$$ Solving for v and noting that $\ell_1 + \ell_2 = 22$ m,

$$v = c\sqrt{\frac{n\lambda}{\ell_1 + \ell_2}} = (3.00\times10^8 \text{ m/s})\sqrt{\frac{(0.005)(589\times10^{-9}\text{ m})}{22\text{ m}}} = 3.47 \text{ km/s}.$$

7. Letting $\ell_1 \to \ell_1\sqrt{1-\beta^2}$ (where $\beta = v/c$) the text equation for t_1 (not currently numbered) becomes

$$t_1 = \frac{2\ell_1\sqrt{1-\beta^2}}{c(1-\beta)} = \frac{2\ell_1}{c\sqrt{1-\beta^2}}$$

which is identical to t_2 when $\ell_1 = \ell_2$, so $\Delta t = 0$ as required.

16. There is no motion in the transverse direction, so $y = z = 3.5$ m.

$$\gamma = \frac{1}{\sqrt{1-\beta^2}} = \frac{1}{\sqrt{1-0.8^2}} = 5/3$$

$$x = \gamma(x' + vt') = \frac{5}{3}(2\text{m} + 0.8c(0)) = 10/3 \text{ m}$$

$$t = \gamma(t' + vx'/c^2) = \frac{5}{3}(0 + (0.8c)(2\text{ m})/c^2) = 8.9\times10^{-9}\text{ s}$$

21. In the muon's frame $T_0 = 2.2$ μs. In the lab frame the time is longer; see Equation (2.19): $T' = \gamma T_0$. In the lab the distance traveled is $9.5\text{cm} = vT' = v\gamma T_0 = \beta c\gamma T_0$, since $v = \beta c$.

Therefore $\beta = \frac{9.5\text{ cm}\left(\sqrt{1-\beta^2}\right)}{cT_0}$, so $\beta = \frac{v}{c} = \frac{9.5\text{ cm}\left(\sqrt{1-\beta^2}\right)}{c(2.2\mu\text{s})}$. Now all quantities are known except β. Solving for β we find $\beta = 1.4\times10^{-4}$ or $v = 1.4\times10^{-4}c$.

26. (a) $L' = L/\gamma = L\sqrt{1-v^2/c^2} = (3.58\times10^4\text{ km})\sqrt{1-0.94^2} = 1.22\times10^4$ km.

(b) Earth's frame: $t = L/v = \frac{3.58\times10^7\text{ m}}{(0.94)(3.00\times10^8\text{ m/s})} = 0.127\text{ s}$

Golf ball's frame: $t' = t/\gamma = 0.127\text{ s}\sqrt{1-0.94^2} = 0.0433\text{ s}$

© 2013 Cengage Learning. All Rights Reserved. May not be scanned, copied or duplicated, or posted to a publicly accessible website, in whole or in part.

31. Start from the formula for velocity addition, Equation (2.23a): $u_x = \dfrac{u'_x + v}{1 + vu'_x / c^2}$.

(a) $u_x = \dfrac{0.62c + 0.84c}{1 + (0.62c)(0.84c)/c^2} = \dfrac{1.46c}{1.52} = 0.96c$

(b) $u_x = \dfrac{-0.62c + 0.84c}{1 + (-0.62c)(0.84c)/c^2} = \dfrac{0.22c}{0.48} = 0.46c$

36. We can ignore the 400 km, which is small compared with the Earth-to-moon distance 3.84×10^8 m. The rotation rate is $\omega = 2\pi \text{ rad} \times 100 \text{ s}^{-1} = 2\pi \times 10^2$ rad/s. Then the speed across the moon's surface is $v = \omega R = \left(2\pi \times 10^2 \text{ rad/s}\right)\left(3.84 \times 10^8 \text{ m}\right) = 2.41 \times 10^{11}$ m/s.

37. Classical: $t = \dfrac{4205 \text{ m}}{0.98c} = 1.43 \times 10^{-5}$ s

Then $N = N_0 \exp\left[\dfrac{-(\ln 2)t}{t_{1/2}}\right] = 14.6$ or about 15 muons.

Relativistic: $t' = t/\gamma = \dfrac{1.43 \times 10^{-5} \text{ s}}{5} = 2.86 \times 10^{-6}$ s

$N = N_0 \exp\left[\dfrac{-(\ln 2)t}{t_{1/2}}\right] = 2710$ muons.

Because of the exponential nature of the decay curve, a factor of five (shorter) in time results in many more muons surviving.

42. For a timelike interval $\Delta s^2 < 0$ so $\Delta x^2 < c^2 \Delta t^2$. We will prove by contradiction. Suppose that there is a frame K′ is which the two events were simultaneous, so that $\Delta t' = 0$. Then by the spacetime invariant $\Delta x^2 - c^2 \Delta t^2 = \Delta x'^2 - c^2 \Delta t'^2 = \Delta x'^2$. But because $\Delta x^2 < c^2 \Delta t^2$, this implies $\Delta x'^2 < 0$, which is impossible because $\Delta x'$ is real.

48. The Doppler shift gives $\lambda = \lambda_0 \sqrt{\dfrac{1-\beta}{1+\beta}}$. With numerical values $\lambda_0 = 650$ nm and $\lambda = 540$ nm, solving this equation for β gives $\beta = 0.183$. The astronaut's speed is $v = \beta c = 5.50 \times 10^7$ m/s. In addition to a red light violation, the astronaut gets a speeding ticket.

51. $f = f_0 \sqrt{\dfrac{1-\beta}{1+\beta}} = (1400 \text{ kHz}) \sqrt{\dfrac{1-0.95}{1+0.95}} = 224$ kHz.

© 2013 Cengage Learning. All Rights Reserved. May not be scanned, copied or duplicated, or posted to a publicly accessible website, in whole or in part.

54. The Doppler shift to higher wavelengths is (with $\lambda_0 = 589$ nm) $\lambda = 700 \text{ nm} = \lambda_0 \sqrt{\dfrac{1+\beta}{1-\beta}}$.

Solving for β we find $\beta = 0.171$. Then $t = \dfrac{v}{a} = \dfrac{(0.171)(3.00\times10^8 \text{ m/s})}{29.4 \text{ m/s}^2} = 1.75\times10^6 \text{ s}$, which is 20.25 days. One problem with this analysis is that we have only computed the time as measured by Earth. We are not prepared to handle the non-inertial frame of the spaceship.

55. Let the instantaneous momentum be in the x-direction and the force be in the y-direction. Then $d\vec{p} = \vec{F}dt = \gamma m d\vec{v}$ and $d\vec{v}$ is also in the y-direction. So we have $\vec{F} = \gamma m \dfrac{d\vec{v}}{dt} = \gamma m \vec{a}$.

60. The initial momentum is $p_0 = \gamma m v = \dfrac{1}{\sqrt{1-(0.5)^2}} m(0.5c) = 0.57735mc$.

(a) $p/p_0 = 1.01 = \dfrac{\gamma m v}{0.57735mc}$ $\qquad \gamma v = (1.01)(0.57735c) = 0.58312c$

Substituting for γ and solving for v, $v = \left[\dfrac{1}{(.58312c)^2} + \dfrac{1}{c^2}\right]^{-1/2} = 0.504c$.

(b) Similarly $v = \left[\dfrac{1}{(.63509c)^2} + \dfrac{1}{c^2}\right]^{-1/2} = 0.536c$

(c) Similarly $v = \left[\dfrac{1}{(1.1547c)^2} + \dfrac{1}{c^2}\right]^{-1/2} = 0.756c$

67. (a) $p = \gamma m u = \dfrac{(511 \text{ keV}/c^2)(0.020c)}{\sqrt{1-0.020^2}} = 10.22 \text{ keV}/c$

$E = \gamma mc^2 = \dfrac{(511 \text{ keV}/c^2)(c^2)}{\sqrt{1-0.02^2}} = 511.102 \text{ keV}$

$K = E - E_0 = 511.102 \text{ keV} - 511.00 \text{ keV} = 102 \text{ eV}$

The results for (b) and (c) follow with similar computations and are tabulated:

β	p (keV/c)	E (keV)	K (keV)
0.20	104.3	521.5	10.5
0.90	1055	1172	661

© 2013 Cengage Learning. All Rights Reserved. May not be scanned, copied or duplicated, or posted to a publicly accessible website, in whole or in part.

68. $E = 2E_0 = \gamma E_0$ so $\gamma = 2$. Then $\beta = \sqrt{1 - \frac{1}{\gamma^2}} = \frac{\sqrt{3}}{2}$ and $v = \frac{\sqrt{3}c}{2}$.

74. The speed is the same for protons, electrons, or any particle.

$$K = (\gamma - 1)mc^2 = 1.01\left(\frac{1}{2}mv^2\right) = 0.505mc^2\beta^2 \text{ so } \gamma - 1 = \frac{1}{\sqrt{1-\beta^2}} - 1 = 0.505\beta^2.$$

Rearranging and solving for β, we find $\beta = 0.114$ or $v = 0.114c$.

77. Up to Equation (2.57) the derivation in the text is complete. Then using the integration by parts formula, $\int x dy = xy - \int y dx$ and noting that in this case $x = u$ and $y = \gamma u$, we have $\int u d(\gamma u) = \gamma u^2 - \int \gamma u du$. Thus

$$K = m\int_0^{\gamma u} u d(\gamma u) = \gamma m u^2 - m\int \gamma u du$$

$$= \gamma m u^2 - m\int \frac{u}{\sqrt{1-u^2/c^2}} du$$

Using integral tables or simple substitution:

$$K = \gamma m u^2 + mc^2\sqrt{1-u^2/c^2}\Big|_0^u$$

$$= \gamma m u^2 + mc^2\sqrt{1-u^2/c^2} - mc^2$$

$$= \frac{mc^2 + mc^2\left(1-u^2/c^2\right)}{\sqrt{1-u^2/c^2}} - mc^2$$

$$= \gamma mc^2 - mc^2 = mc^2(\gamma - 1)$$

81. $E = K + E_0 = 1 \text{ TeV} + 938 \text{ MeV} \approx 1 \text{ TeV}$

$$p = \frac{\sqrt{E^2 - E_0^2}}{c} = \frac{\sqrt{(1 \text{ TeV} + 938\text{MeV})^2 - (938 \text{ MeV})^2}}{c} = 1.000938 \text{ TeV}/c$$

$$\gamma = \frac{E + E_0}{E_0} = \frac{1.000938 \text{ TeV}}{0.000938 \text{ TeV}} = 1067 \qquad \beta^2 = 1 - \frac{1}{\gamma^2} = 1 - 8.78\times10^{-7}$$

$$\beta = \sqrt{1 - 8.78\times10^{-7}} \approx 1 - 4.39\times10^{-7} \text{ and } v = \beta c \approx 0.999999561c$$

83. $E = K + E_0 = 200\text{MeV} + 106\text{MeV} = 306 \text{ MeV}$

$$p = \frac{\sqrt{E^2 - E_0^2}}{c} = \frac{\sqrt{(306 \text{ MeV})^2 - (106 \text{ MeV})^2}}{c} = 287.05 \text{ MeV}/c$$

© 2013 Cengage Learning. All Rights Reserved. May not be scanned, copied or duplicated, or posted to a publicly accessible website, in whole or in part.

$$\gamma = \frac{E}{E_0} = \frac{306\text{ MeV}}{106\text{ MeV}} = 2.887$$

$$\beta = \sqrt{1 - \frac{1}{\gamma^2}} = 0.938$$

so $v = 0.938c$

85. (a) The mass-energy imbalance occurs because the helium-4 (^{4}He) is more tightly bound than the deuterium (^{2}H) and tritium nuclei (^{3}H). (Masses from Appendix 8.)

$$\Delta E = \left([m(^2\text{H}) + m(^3\text{H})] - [m_n + m(^4\text{He})]\right)c^2$$

$$= [(2.014102\text{u} + 3.016029\text{u}) - (1.008665\text{u} + 4.002603\text{u})]c^2\left(\frac{931.494\text{ MeV}}{c^2 \cdot \text{u}}\right)$$

$$= 17.6\text{ MeV}$$

(b) The initial rest energy is

$\left[m(^2\text{H}) + m(^3\text{H})\right] \cdot c^2 = [(5.030131\text{ u})] \cdot c^2\left(\frac{931.494\text{ MeV}}{c^2 \cdot \text{u}}\right)$4686 MeV. Thus the answer in (a) is about 0.37% of the initial rest energy.

87. As in the solution to Problem 21 we have $\beta = \frac{v}{c} = \frac{d\sqrt{1-\beta^2}}{ct'}$ where d is the length of the particle track and t' the particle's lifetime in its rest frame. In this problem $t' = 8.2 \times 10^{-11}$ s and d = 24 mm. Solving the above equation we find $\beta = 0.698$. Then

$$E = \frac{E_0}{\sqrt{1-\beta^2}} = \frac{1672\text{ MeV}}{\sqrt{1-0.698^2}} = 2330\text{ MeV}.$$

89. (a) The number n received by Frank at f' is half the number sent by Mary at that rate, or $fL/\gamma v$. The detected time of turnaround is

$$t = \frac{n}{f'} = \frac{fL/\gamma v}{v\sqrt{(1-\beta)/(1+\beta)}} = \frac{L\sqrt{1+\beta}}{\gamma v\sqrt{1-\beta}} = \frac{L(1+\beta)}{v} = \frac{L}{v} + \frac{L}{c}.$$

(b) Similarly, the number n' received by Mary at f' is

$$n' = f'\frac{T'}{2} = f\sqrt{\frac{1-\beta}{1+\beta}}\frac{L}{\gamma v} = \frac{f L(1-\beta)}{v}.$$

Her turnaround time is $T'/2 = L/\gamma v$.

(c) For Frank, the time t_2 for the remainder of the trip is $t_2 = T - t_1 = L/v - L/c$.

© 2013 Cengage Learning. All Rights Reserved. May not be scanned, copied or duplicated, or posted to a publicly accessible website, in whole or in part.

Number of signals = $f''t_2 = f\sqrt{\frac{1+\beta}{1-\beta}}(L/v - L/c) = \frac{fL}{\gamma v}$.

Total number received = $\frac{fL}{\gamma v} + \frac{fL}{\gamma v} = \frac{2fL}{\gamma v}$.

Mary's age = $\frac{\text{Total number received}}{f} = \frac{2L}{\gamma v}$.

(d) For Mary, $t_2' = T' - t_1' = L/\gamma v$.

Number of signals = $f''t_2' = f\sqrt{\frac{1+\beta}{1-\beta}}\frac{L}{\gamma v} = \frac{fL}{v}(1+\beta)$.

Total number received = $\frac{fL}{v}(1-\beta+1+\beta) = \frac{2fL}{v}$.

Frank's age = $\frac{\text{Total number received}}{f} = \frac{2L}{v}$.

95. (a) For the proton: $p = \gamma mu = \frac{1}{\sqrt{1-0.9^2}}(938\text{ MeV}/c^2)(0.9c) = 1940\text{ MeV}/c$.

For the electron: $E = \sqrt{p^2c^2 + E_0^2} = \sqrt{(1940\text{ MeV})^2 + (0.511\text{ MeV})^2} = 1940\text{ MeV}$.

$$\gamma = \frac{E}{E_0} = \frac{1940\text{ MeV}}{0.511\text{ MeV}} = 3797$$

$$\beta = \sqrt{1-\frac{1}{\gamma^2}} = \sqrt{1-\frac{1}{3797^2}} = \sqrt{1-6.94\times10^{-8}} \approx 1-3.97\times10^{-8}$$

$v = (1-3.97\times10^{-8})c.$

(b) For the proton: $K = (\gamma-1)E_0 = \left(\frac{1}{\sqrt{1-0.9^2}} - 1\right)(938\text{ MeV}) = 1214\text{ MeV}$.

For the electron: $\gamma = \frac{K+E_0}{E_0} = \frac{1214\text{ MeV} + 0.511\text{ MeV}}{0.511\text{ MeV}} = 2377$.

$$\beta = \sqrt{1-\frac{1}{\gamma^2}} = \sqrt{1-\frac{1}{2377^2}} = \sqrt{1-1.77\times10^{-7}} \approx 1-8.85\times10^{-8}$$

$v = (1-8.85\times10^{-8})c.$

102. As we know that the quasars are moving away at high speeds, we make use of Equation (2.33) and the equation $c = \lambda f$. Using a prime to indicate the Doppler shifted frequency

© 2013 Cengage Learning. All Rights Reserved. May not be scanned, copied or duplicated, or posted to a publicly accessible website, in whole or in part.

(or wavelength), Equation (2.33) indicates that frequency is given by $f' = \frac{\sqrt{1-\beta}}{\sqrt{1+\beta}} f_0$, or

$$\frac{f_0}{f'} = \frac{\sqrt{1+\beta}}{\sqrt{1-\beta}} \text{ so}$$

$$z = \frac{(\lambda' - \lambda_0)}{\lambda_0} = \left(\frac{\lambda'}{\lambda_0} - 1\right) = \left(\frac{c/f'}{c/f_0} - 1\right)$$

$$= \left(\frac{f_0}{f'} - 1\right) = \left(\frac{\sqrt{1+\beta}}{\sqrt{1-\beta}} - 1\right)$$

Therefore $(z+1)^2 = \frac{1+\beta}{1-\beta}$. We can complete the algebra to show that

$\beta = \frac{(z+1)^2 - 1}{(z+1)^2 + 1}$ and thus $v = \left[\frac{(z+1)^2 - 1}{(z+1)^2 + 1}\right] c$. For the values of z given, $v = 0.787c$ for $z = 1.9$ and $v = 0.944c$ for $z = 4.9$.

103. (a) In the frame of the decaying K^0 meson, the pi mesons must recoil with equal momenta in opposite directions in order to conserve momentum. In that reference frame the available kinetic energy is $498 \text{ MeV} - 2(135 \text{ MeV}) = 228 \text{ MeV}$.

(b) The pi mesons share this equally, so each one has a kinetic energy of 114 MeV in that frame. The energy of each pi meson is $E = K + E_0 = 114 \text{ MeV} + 135 \text{ MeV} = 249 \text{ MeV}$. The momentum of each pi meson can be found:

$$p = \frac{\sqrt{E^2 - E_0^2}}{c} = \frac{\sqrt{(249 \text{ MeV})^2 - (135 \text{ MeV})^2}}{c} = 209.23 \text{ MeV}/c.$$

© 2013 Cengage Learning. All Rights Reserved. May not be scanned, copied or duplicated, or posted to a publicly accessible website, in whole or in part.

Chapter 3

4. $eE = evB$ so $E = vB = \left(4.0\times10^{6}\ \text{m/s}\right)\left(1.2\times10^{-2}\ \text{T}\right) = 4.8\times10^{4}\,\text{V/m}.$

$$y = \frac{1}{2}at^2 = \frac{1}{2}\left(\frac{F}{m}\right)\left(\frac{\ell}{v_0}\right)^2 = \frac{1}{2}\left(\frac{eE}{m}\right)\left(\frac{\ell}{v_0}\right)^2 = \frac{eE\ell^2}{2mv_0^2}$$

$$= \frac{\left(1.602\times10^{-19}\ \text{C}\right)\left(4.8\times10^{4}\ \text{V/m}\right)\left(0.02\ \text{m}\right)^2}{2\left(9.109\times10^{-31}\ \text{kg}\right)\left(4.0\times10^{6}\ \text{m/s}\right)^2} = 1.0552\times10^{-1}\ \text{m} = 10.6\ \text{cm}.$$

7. $v_t = \dfrac{mg}{f} = \dfrac{mg}{6\pi\eta r}$ $\qquad m = \rho(\text{volume}) = \dfrac{4}{3}\pi\rho r^3$

$$v_t = \left(\frac{4}{3}\pi\rho r^3\right)\left(\frac{g}{6\pi\eta r}\right) = \frac{2g\rho r^2}{9\eta}$$

Solving for r: $r = 3\sqrt{\dfrac{\eta v_t}{2g\rho}}$

9. Lyman: $\lambda = \left[R_H\left(1 - \dfrac{1}{\infty^2}\right)\right]^{-1} = R_H^{-1} = \left(1.096776\times10^{7}\ \text{m}^{-1}\right)^{-1} = 91.2\ \text{nm}$

Balmer: $\lambda = \left[R_H\left(\dfrac{1}{2^2} - \dfrac{1}{\infty^2}\right)\right]^{-1} = 4R_H^{-1} = 4\left(1.096776\times10^{7}\ \text{m}^{-1}\right)^{-1} = 364.7\ \text{nm}$

13. Beginning with Equation (3.10) with $n = 1$, we have $\lambda = d\sin\theta$ with $d = 0.20\text{mm}$. Therefore $\dfrac{d\lambda}{d\theta} = d\cos\theta$. Assuming the spectrum was viewed in the forward direction, then $\theta \approx 0$ and $\cos\theta \approx 1$ so $\dfrac{d\lambda}{d\theta} \approx \dfrac{\Delta\lambda}{\Delta\theta} = d$. An angle of 0.50 minutes of arc corresponds to 1.45×10^{-4} rad, so

$$\Delta\lambda = d\Delta\theta = \left(0.20\times10^{-3}\text{m}\right)1.45\times10^{-4}\,\text{rad} = 29\ \text{nm}.$$

19. (a) $\dfrac{P_1}{P_0} = \dfrac{\sigma T_1^4}{\sigma T_0^4}$; so $P_1 = P_0\dfrac{T_1^4}{T_0^4} = \left(\dfrac{2300\ \text{K}}{900\ \text{K}}\right)^4 P_0 = 42.7\ P_0$; The power increases by a factor of 42.7.

(b) To double the power output, the ratio of temperatures to the fourth power must equal 2: $\left(\dfrac{T_1}{900\ \text{K}}\right)^4 = 2$. Solving we find $T_1 = 1070\ \text{K}$.

© 2013 Cengage Learning. All Rights Reserved. May not be scanned, copied or duplicated, or posted to a publicly accessible website, in whole or in part.

20. (a) $\lambda_{max} = \dfrac{2.898\times10^{-3}\ \text{m}\cdot\text{K}}{310\ \text{K}} = 9.35\ \mu\text{m}.$

(b) At this temperature the power per unit area is
$R = \sigma T^4 = \left(5.67\times10^{-8}\,\text{W}\cdot\text{m}^{-2}\cdot\text{K}^4\right)(310\text{K})^4 = 524\ \text{W/m}^2$. The total surface area of a cylinder is $2\pi r(r+h) = 2\pi(0.13\ \text{m})(1.75\ \text{m}+0.13\ \text{m}) = 1.54\ \text{m}^2$ so the total power is $P = \left(524\ \text{W/m}^2\right)\left(1.54\ \text{m}^2\right) = 807\ \text{W}.$

(c) The total energy radiated in one day is the power multiplied by the time:
$E = P\cdot t = (807\ \text{W})\cdot(86400\ \text{s}) = 6.97\times10^7\,\text{J}.$

2000 kcal $= \left(2\times10^6\ \text{cal}\right)\cdot(4.186\,\text{J/cal}) = 8.37\times10^6\ \text{J}$.

There are several assumptions. First, a cylinder may overestimate the total surface area; second, radiation is minimized by hair covering and clothing.

23. $\lambda_{max} = \dfrac{2.898\times10^{-3}\ \text{m}\cdot\text{K}}{3000\ \text{K}} = 966\ \text{nm}$ which is in the near infrared.

28. Taking derivatives $\dfrac{\partial^2\psi}{\partial t^2} = \dfrac{\partial^2 a}{\partial t^2}\sin\left(\dfrac{n\pi x}{L}\right)$; $\dfrac{\partial^2\psi}{\partial x^2} = -a\dfrac{n^2\pi^2}{L^2}\sin\left(\dfrac{n\pi x}{L}\right)$. Substituting these values into the wave equation produces $\dfrac{1}{c^2}\dfrac{\partial^2 a}{\partial t^2}\sin\left(\dfrac{n\pi x}{L}\right) - \left(-a\dfrac{n^2\pi^2}{L^2}\sin\left(\dfrac{n\pi x}{L}\right)\right) = 0$ which simplifies to $\dfrac{\partial^2 a}{\partial t^2} = -a\dfrac{n^2\pi^2 c^2}{L^2} = -\omega^2 a$ where $\omega = \dfrac{n\pi c}{L}$. Because $\lambda = \dfrac{2L}{n}$ for this system and $c = \lambda f$, then $\omega = 2\pi f$.

32. Energy per photon $= hf = \left(6.626\times10^{-34}\,\text{J}\cdot\text{s}\right)\left(98.1\times10^6\text{s}^{-1}\right) = 6.50\times10^{-26}$ J.

$\left(5.0\times10^4\ \text{J/s}\right)\dfrac{1\ \text{photon}}{6.50\times10^{-26}\ \text{J}} = 7.69\times10^{29}$ photons/s

35. $\lambda_t = \dfrac{hc}{\phi} = \dfrac{1240\ \text{eV}\cdot\text{nm}}{4.64\ \text{eV}} = 267.2\ \text{nm}.$

If the wavelength is halved (to $\lambda = 133.6\,\text{nm}$), then

$$K = \frac{hc}{\lambda} - \phi = \frac{1240\ \text{eV}\cdot\text{nm}}{133.6\ \text{nm}} - 4.64\ \text{eV}\ = 4.64\ \text{eV}.$$

© 2013 Cengage Learning. All Rights Reserved. May not be scanned, copied or duplicated, or posted to a publicly accessible website, in whole or in part.

43. $\lambda = \dfrac{1240 \text{ eV}\cdot\text{nm}}{2.5\times10^4 \text{ eV}} = 0.0496 \text{ nm}.$

45. From Figure (3.19) we observe that the two characteristic spectral lines occur at wavelengths of $6.4\times10^{-11}\text{m}$ and $7.2\times10^{-11}\text{m}$. Rearrange Equation (3.37) to solve for the potential V_0; $V_0 = \dfrac{hc}{e}\dfrac{1}{\lambda_{\text{min}}} = \dfrac{1.24\times10^{-6}\text{V}\cdot\text{m}}{7.2\times10^{-11}\text{m}} = 17.2$ kV and we have used the larger of the two wavelengths in order to determine the minimum potential difference.

50. $\dfrac{\Delta\lambda}{\lambda} = 0.004 = \dfrac{\lambda_c}{\lambda}(1-\cos\theta)$ so $\lambda = 250\lambda_c(1-\cos\theta)$

(a) $\lambda = 250\left(2.43\times10^{-12}\text{ m}\right)\left(1-\cos30^\circ\right) = 8.14\times10^{-11}\text{ m}$

(b) $\lambda = 250\left(2.43\times10^{-12}\text{ m}\right)\left(1-\cos90^\circ\right) = 6.08\times10^{-10}\text{ m}$

(c) $\lambda = 250\left(2.43\times10^{-12}\text{ m}\right)\left(1-\cos170^\circ\right) = 1.21\times10^{-9}\text{ m}$

53. For $\theta = 90^\circ$ we know $\lambda' = \lambda + \lambda_c = 2.00243\text{nm}$. $\dfrac{\Delta\lambda}{\lambda} = \dfrac{\lambda_c}{\lambda} = \dfrac{2.43\times10^{-3}\text{ nm}}{2\text{ nm}} = 1.22\times10^{-3}$.

54. $E = 2mc^2 = 2(938.3\text{ MeV}) = 1877\text{ MeV}$. This energy could come from a particle accelerator.

60. For maximum recoil energy the scattering angle is $\theta = 180^\circ$ and $\phi = 0$. Then as usual $\Delta\lambda = \dfrac{2h}{mc}$. Using the result of Problem 56,

$$K = \frac{\Delta\lambda/\lambda}{1+\Delta\lambda/\lambda}hf = \frac{2h/mc\lambda}{1+2h/mc\lambda}hf = \frac{2hf/mc^2}{1+2hf/mc^2}hf.$$

We can rearrange this equation: $K\left(1+\dfrac{2hf}{mc^2}\right) = \dfrac{2(hf)^2}{mc^2}$ or

$\left(\dfrac{2}{mc^2}\right)(hf)^2 - \left(\dfrac{2K}{mc^2}\right)(hf) - K = 0$. This constitutes a quadratic equation in hf which can be solved with the given value of $K = 100\text{keV}$ numerically to yield $hf = 217$ keV.

© 2013 Cengage Learning. All Rights Reserved. May not be scanned, copied or duplicated, or posted to a publicly accessible website, in whole or in part.

65. The laser's power is 25 kW = 2.5×10^4 J/s. The energy of a single photon is

$$E = \frac{hc}{\lambda} = \frac{(6.626\times10^{-34}\ \text{J}\cdot\text{s})(3.00\times10^{8}\ \text{m/s})}{1.06\times10^{-6}\ \text{m}} = 1.88\times10^{-19}\ \text{J}.$$ Thus,

$$2.5\times10^{4}\ \text{J/s}\ \frac{1\ \text{photon}}{1.88\times10^{-19}\ \text{J}} = 1.3\times10^{23}\ \text{photons/s}.$$

67. (a) $$\lambda_{\text{max}} = \frac{2.898\times10^{-3}\,\text{m}\cdot\text{K}}{9600\ \text{K}} = 3.01\times10^{-7}\,\text{m} = 301\ \text{nm}.$$

This is actually in the ultraviolet part of the spectrum, but the shape of the distribution function guarantees that most of the visible photons will be in the violet-blue region, which explains the star's blue appearance.

(b) Assuming a perfect blackbody,

$$P = \sigma T^4 A = \left(5.67\times10^{-8}\ \text{W}/\left(\text{m}^2\cdot\text{K}^4\right)\right)(9600\ \text{K})^4\left[4\pi\left(1.6\times6.96\times10^{8}\ \text{m}\right)^2\right]$$

$$= 7.51\times10^{27}\ \text{W}.$$

This is about 19 times the value for the sun, reported in Example 3.5.

© 2013 Cengage Learning. All Rights Reserved. May not be scanned, copied or duplicated, or posted to a publicly accessible website, in whole or in part.

Chapter 4

5. (a) With $Z_1 = 2$, $Z_2 = 79$, and $\theta = 1^\circ$ we have

$$b = \frac{Z_1 Z_2 e^2}{8\pi\epsilon_0 K}\cot\left(\frac{\theta}{2}\right) = \frac{(2)(79)\left(1.44\times10^{-9}\text{ eV}\cdot\text{m}\right)}{2\left(7.7\times10^{6}\text{ eV}\right)}\cot\left(0.5^\circ\right) = 1.69\times10^{-12}\text{ m}$$

(b) For $\theta = 90^\circ$ $b = \frac{Z_1 Z_2 e^2}{8\pi\epsilon_0 K}\cot\left(\frac{\theta}{2}\right) = \frac{(2)(79)\left(1.44\times10^{-9}\text{ eV}\cdot\text{m}\right)}{2\left(7.7\times10^{6}\text{ eV}\right)}\cot\left(45^\circ\right) = 1.48\times10^{-14}\text{ m}$

6. $f = \pi n t\left(\frac{Z_1 Z_2 e^2}{8\pi\epsilon_0 K}\right)^2\cot^2\left(\frac{\theta}{2}\right)$. For the two different angles everything is the same except the angles, so the ratio is $\frac{f(1^\circ)}{f(2^\circ)} = \frac{\cot^2(0.5^\circ)}{\cot^2\left(1.0^\circ\right)} = 4.00.$

10. From the Rutherford scattering result, the number detected through a small angle is inversely proportional to $\sin^4\left(\frac{\theta}{2}\right)$. Thus $\frac{n(50^\circ)}{n(6^\circ)} = \frac{\sin^4(3^\circ)}{\sin^4\left(25^\circ\right)} = 2.35\times10^{-4}$ and if they count 2000 at 6° the number counted at 50° will be $(2000)\left(2.35\times10^{-4}\right) = 0.47$, which is insufficient.

15. (a) $v = \frac{e}{\sqrt{4\pi\epsilon_0 m r}} = \frac{ec}{\sqrt{4\pi\epsilon_0 m c^2 r}} = \frac{\sqrt{1.44\text{ eV}\cdot\text{nm}}}{\sqrt{\left(938\times10^{6}\text{ eV}\right)(0.05\text{ nm})}}c = 1.75\times10^{-4}c$

or $v = 5.25\times10^{4}$ m/s .

(b) $E = -\frac{e^2}{8\pi\epsilon_0 r} = -\frac{1.44\text{ eV}\cdot\text{nm}}{2(0.05\text{ nm})} = -14.4\text{ eV}.$

(c) The "nucleus"' is too light to be fixed, and there is no way to reconcile this model with the results of Rutherford scattering.

23. The photon energy is $E = \frac{hc}{\lambda} = \frac{1240\text{ eV}\cdot\text{nm}}{410\text{ nm}} = 3.02\text{ eV}$. This is the energy difference between the two states in hydrogen. From Figure 4.16, we see $E_3 = -1.51$ eV so the initial state must be $n = 2$ in order to have a large enough difference. This energy difference exists between $n = 2$ (with $E_2 = -3.40$ eV) and $n = 6$ (with $E_6 = -0.38$ eV).

© 2013 Cengage Learning. All Rights Reserved. May not be scanned, copied or duplicated, or posted to a publicly accessible website, in whole or in part.

29. From Equation (4.31), $v_n = (1/n)(\hbar / ma_0)$which gives the speed in the $n = 3$ state: $v = 7.30\times10^5$ m/s. The radius of the orbit is $n^2 a_0 = 9a_0$. Then from kinematics:

$$\text{number of revolutions } = \frac{vt}{2\pi r} = \frac{\left(7.30\times10^5 \text{ m/s}\right)\left(10^{-8} \text{ s}\right)}{2\pi(9)\left(5.29\times10^{-11} \text{ m}\right)} = 2.44\times10^6.$$

30. The energy of each photon is $hc/\lambda = \left(\frac{1239.8 \text{ eV}\cdot\text{nm}}{397\,\text{nm}}\right) = 3.12$ eV. Looking at the energy difference between levels in hydrogen (with $E_n = -E_0/n^2$) we see that $E_7 - E_2 = 3.12$ eV, which matches the photon energy to three significant figures. Therefore this laser can promote the atom to the $n = 7$ level.

31. We must use the reduced mass for the muon:

$$\mu = \frac{mM}{m+M} = \frac{\left(106 \text{ MeV/}c^2\right)\left(938 \text{ MeV/}c^2\right)}{106 \text{ MeV/}c^2 + 938 \text{ MeV}/c^2} = 95.2 \text{ MeV/c}^2$$

(a) $$a_0 = \frac{4\pi\epsilon_0\hbar^2}{\mu e^2} = \frac{\left(6.58\times10^{-16} \text{ eV}\cdot\text{s}\right)^2}{\left(1.44\times10^{-9} \text{ eV}\cdot\text{m}\right)\left(95.2\times10^6 \text{ eV/}c^2\right)} \frac{\left(3.00\times10^8 \text{ m/s}\right)^2}{c^2} = 2.84\times10^{-13} \text{ m}$$

(b) $$E = \frac{e^2}{8\pi\epsilon_0 a_0} = \frac{\left(1.44\times10^{-9} \text{ eV}\cdot\text{m}\right)}{2\left(2.84\times10^{-13} \text{ m}\right)} = 2535 \text{ eV}$$

(c) First series: $\lambda = \frac{hc}{E} = \frac{1240 \text{ eV}\cdot\text{nm}}{2535 \text{ eV}} = 0.49$ nm

Second series: $\lambda = \frac{4hc}{E} = \frac{4(1240 \text{ eV}\cdot\text{nm})}{2535 \text{ eV}} = 1.96$ nm

Third series: $\lambda = \frac{9hc}{E} = \frac{9(1240 \text{ eV}\cdot\text{nm})}{2535 \text{ eV}} = 4.40$ nm

38. For L_α we have $\lambda = \frac{c}{f} = \frac{36}{5R(Z-7.4)^2}$.

$$Z = 43 : \lambda = \frac{36}{5R(43\text{-}7.4)^2} = 0.52 \text{ nm};$$

$$Z = 61 : \lambda = \frac{36}{5R(61-7.4)^2} = 0.23 \text{ nm};$$

$$Z = 75 : \lambda = \frac{36}{5R(75-7.4)^2} = 0.14 \text{ nm}.$$

© 2013 Cengage Learning. All Rights Reserved. May not be scanned, copied or duplicated, or posted to a publicly accessible website, in whole or in part.

41. Helium: $\lambda(\mathrm{K}_\alpha)=\dfrac{4}{3R(Z-1)^2}=122\,\mathrm{nm}$; $\lambda(\mathrm{K}_\beta)=\dfrac{9}{8R(Z-1)^2}=103\ \mathrm{nm}$

Lithium: $\lambda(\mathrm{K}_\alpha)=\dfrac{4}{3R(Z-1)^2}=30.4\,\mathrm{nm}$; $\lambda(\mathrm{K}_\beta)=\dfrac{9}{8R(Z-1)^2}=25.6\ \mathrm{nm}$

44. The longest wavelengths occur for an electron vacancy in K shell. The two longest wavelengths correspond to the K_α and the K_β. Use Equation (4.41) and rearrange it to solve for $(Z-1)^2$. Then

$$(Z-1)^2=\frac{4}{3}\frac{1}{R\lambda_{\mathrm{K}_\alpha}}=\frac{4}{3\left(1.09737\times10^7\,\mathrm{m}^{-1}\right)\left(0.155\times10^{-9}\,\mathrm{m}\right)}=783.89.$$

This gives $(Z-1)=28$ or $Z=29$.

Using the expression for λ_{K_β} from Problem 40, we have $(Z-1)^2=\dfrac{9}{8}\dfrac{1}{R\lambda_{\mathrm{K}_\beta}}$.

Using the second wavelength given, 0.131 nm, we find

$$(Z-1)^2=\frac{9}{8}\frac{1}{R\lambda_{\mathrm{K}_\beta}}=\frac{9}{8}\frac{1}{\left(1.09737\times10^7\,\mathrm{m}^{-1}\right)\left(0.131\times10^{-9}\,\mathrm{m}\right)}=782.58.$$

This yields $(Z-1)=27.97$ or $Z=29$. Therefore we can conclude the target must be copper.

50. K_α is a transition from $n=2$ to $n=1$ and K_β is from $n=3$ to $n=1$. We know those wavelengths in the Lyman series are 121.6 nm and 102.6 nm, respectively. The redshift factor (λ/λ_0) is (with $\beta=1/6$) $\sqrt{\dfrac{1+\beta}{1-\beta}}=\sqrt{\dfrac{1+1/6}{1-1/6}}=1.183$. Then the redshifted wavelengths are higher by 18.3% in each case. The observed wavelengths are:
K_α: $\lambda=(1.183)(121.6\ \mathrm{nm})=143.9\ \mathrm{nm}$ and K_β: $\lambda=(1.183)(102.6\ \mathrm{nm})=121.4\ \mathrm{nm}$.

51. $f=\pi nt\left(\dfrac{Z_1Z_2e^2}{8\pi\epsilon_0K}\right)^2\cot^2\left(\dfrac{\theta}{2}\right)$

(a) $$f(1^\circ)=\pi\left(5.90\times10^{28}\ \mathrm{m}^{-3}\right)\left(0.32\times10^{-6}\ \mathrm{m}\right)\left(\frac{(2)(79)\left(1.44\times10^{-9}\ \mathrm{eV\cdot m}\right)}{2\left(8\times10^6\ \mathrm{eV}\right)}\right)^2\cot^2\left(0.5^\circ\right)$$

$=0.157$

$$f(2^\circ)=\pi\left(5.90\times10^{28}\ \mathrm{m}^{-3}\right)\left(0.32\times10^{-6}\ \mathrm{m}\right)\left(\frac{(2)(79)\left(1.44\times10^{-9}\ \mathrm{eV\cdot m}\right)}{2\left(8\times10^6\ \mathrm{eV}\right)}\right)^2\cot^2\left(1^\circ\right)$$

$=0.0394$

© 2013 Cengage Learning. All Rights Reserved. May not be scanned, copied or duplicated, or posted to a publicly accessible website, in whole or in part.

The fraction scattered between 1° and 2° is $0.157 - 0.0394 = 0.118$.

(b) $\dfrac{f(1^\circ)}{f(10^\circ)} = \dfrac{\cot^2(0.5^\circ)}{\cot^2(5^\circ)} = 100.5$; $\dfrac{f(1^\circ)}{f(90^\circ)} = \dfrac{\cot^2(0.5^\circ)}{\cot^2(45^\circ)} = 1.31\times10^4$

55. We start with $K = nhf_{\text{orb}}/2$. From classical mechanics we have for a circular orbit

$f = v/2\pi r$, or $r = v/2\pi f$: $L = mvr = mv\left(\dfrac{v}{2\pi f}\right) = \left(\dfrac{mv^2}{2}\right)\left(\dfrac{1}{\pi f}\right)$.

Using $K = \dfrac{1}{2}mv^2$, $L = \dfrac{K}{\pi f} = \dfrac{nhf}{2\pi f} = \dfrac{nh}{2\pi} = n\hbar$.

57. The longest wavelength in the Lyman series is 121.57 nm, using the value of R_H. Then from the Doppler-shift formula, $\dfrac{\lambda}{\lambda_0} = \dfrac{137.15\,\text{nm}}{121.57\,\text{nm}} = \sqrt{\dfrac{1+\beta}{1-\beta}}$. We can solve to find $\beta = 0.12$ or $v = 3.6\times10^7$ m/s.

© 2013 Cengage Learning. All Rights Reserved. May not be scanned, copied or duplicated, or posted to a publicly accessible website, in whole or in part.

Chapter 5

2. Use $\lambda = 0.207\,\text{nm}$, and we know from the text (Example 5.1) that $d = 0.282$ nm for NaCl.

$$n=1:\ \sin\theta = \frac{n\lambda}{2d} = \frac{\lambda}{2d} = 0.367;\ \theta = 21.3^\circ$$

$$n=2:\ \sin\theta = \frac{n\lambda}{2d} = \frac{\lambda}{d} = 0.734;\ \theta = 47.2^\circ$$

$$\Delta\lambda = 47.2^\circ - 21.3^\circ = 25.9^\circ$$

8. The resolution will be comparable to the de Broglie wavelength. The energy of the microscope requires a relativistic treatment, so

$$\lambda = \frac{h}{p} = \frac{hc}{\sqrt{K^2 + 2\left(mc^2\right)K}}$$

$$= \frac{1240\text{eV}\cdot\text{nm}}{\sqrt{\left(3.0\times10^6\,\text{eV}\right)^2 + 2\left(0.511\times10^6\,\text{eV}\right)\left(3.0\times10^6\,\text{eV}\right)}}$$

$$= 3.57\times10^{-4}\,\text{nm} = 0.357\,\text{pm}$$

11. (a) Relativistically $p = \dfrac{\sqrt{E^2 - E_0^2}}{c} = \dfrac{\sqrt{\left(K+mc^2\right)^2 - \left(mc^2\right)^2}}{c} = \dfrac{\sqrt{K^2 + 2Kmc^2}}{c}$;

$$\lambda = \frac{h}{p} = \frac{hc}{\sqrt{K^2 + 2Kmc^2}}$$

(b) Non-relativistically, as in the text $\lambda = \dfrac{h}{\sqrt{2mK}} = \dfrac{hc}{\sqrt{2mc^2K}}$.

12. The rest energy of the electron is very small compared to the total energy.

$$p = \frac{\sqrt{E^2 - E_0^2}}{c} = 50\ \text{GeV}/c;\ \lambda = \frac{h}{p} = \frac{1240\ \text{eV}\cdot\text{nm}}{50\ \text{GeV}} = 2.48\times10^{-17}\ \text{m};$$

$$\text{fraction} = \frac{2.48\times10^{-17}\ \text{m}}{2\times10^{-15}\,\text{m}} = 0.012.$$

15. From the accelerating potential we know $K = eV = 3.00\,\text{keV}$, so $E = K + E_0 = 514$ keV.

$$p = \frac{\sqrt{E^2 - E_0^2}}{c} = \frac{\sqrt{\left(514\ \text{keV}\right)^2 - \left(511\ \text{keV}\right)^2}}{c} = 55.4\ \text{keV}/c;$$

$$\lambda = \frac{h}{p} = \frac{hc}{pc} = \frac{1240\ \text{eV}\cdot\text{nm}}{55.4\times10^3\ \text{eV}} = 22.4\ \text{pm}.$$

© 2013 Cengage Learning. All Rights Reserved. May not be scanned, copied or duplicated, or posted to a publicly accessible website, in whole or in part.

22. $\lambda = \dfrac{h}{\sqrt{2mK}} = \dfrac{hc}{\sqrt{2mc^2K}} = \dfrac{1240\text{ eV}\cdot\text{nm}}{\sqrt{2\left(939\times10^6\text{ eV}\right)\left(0.025\text{ eV}\right)}} = 0.181\text{ nm};$

$\lambda = D\sin\phi;\ \ \phi = \sin^{-1}\left(\dfrac{\lambda}{D}\right) = \sin^{-1}\left(\dfrac{0.181\text{ nm}}{0.45\text{ nm}}\right) = 23.7^\circ.$

29. As in Example 5.6 we start with $v_{ph} = c\lambda^n$ where c is a constant. We also know from Equation (5.33) that $u_g = v_{ph} + k\dfrac{dv_{ph}}{dk}$. We note further that $\dfrac{dv_{ph}}{dk} = \dfrac{dv_{ph}}{d\lambda}\left(\dfrac{d\lambda}{dk}\right)$. Since $\lambda = \dfrac{2\pi}{k}$ then $\dfrac{d\lambda}{dk} = \dfrac{-2\pi}{k^2} = \dfrac{-\lambda^2}{2\pi}$. Therefore $u_g = v_{ph} + \left(\dfrac{2\pi}{\lambda}\right)\dfrac{dv_{ph}}{d\lambda}\left(\dfrac{-\lambda^2}{2\pi}\right) = v_{ph} - (\lambda)\dfrac{dv_{ph}}{d\lambda}$.

Setting $u_g = v_{ph} = c\lambda^n$, we find $c\lambda^n = c\lambda^n - cn\lambda^n$. This can be satisfied only if $n = 0$, so v_{ph} is independent of λ. This is consistent with the idea that when a medium is non-dispersive, the phase and group velocities are equal and the speed independent of wavelength.

32. Relativistically: $u = \dfrac{dE}{dp} = \dfrac{d}{dp}\left(p^2c^2 + E_0^2\right)^{1/2} = \dfrac{pc^2}{\sqrt{p^2c^2 + E_0^2}} = \dfrac{pc^2}{E};$

Classically: $u = \dfrac{dE}{dp} = \dfrac{d}{dp}\left(\dfrac{p^2}{2m}\right) = \dfrac{p}{m}$

35. We make use of the small angle approximations: $\sin\theta \approx \tan\theta$ and $\sin\theta \approx \theta$.

$\sin\theta \approx \tan\theta = \dfrac{0.3\text{ mm}}{0.8\text{m}} = 3.75\times10^{-4};$

$\lambda = d\sin\theta \approx d\theta = (2000\,\text{nm})\left(3.75\times10^{-4}\right) = 0.75\text{ nm};$

$p = \dfrac{h}{\lambda} = \dfrac{hc}{\lambda c} = \dfrac{1240\text{ eV}\cdot\text{nm}}{(0.75\text{ nm})c} = 1.653\text{ keV}/c;$

$K = E - E_0 = \sqrt{p^2c^2 + E_0^2} - E_0 = \sqrt{(1.653\text{ keV})^2 + (511\text{ keV})^2} - 511\text{ keV} = 2.67\text{eV}.$

Such low energies will present problems, because low-energy electrons take longer to move through the region of the field and thus the deflection by stray electric fields is greater.

© 2013 Cengage Learning. All Rights Reserved. May not be scanned, copied or duplicated, or posted to a publicly accessible website, in whole or in part.

38. The uncertainty ratio is the same for any mass and independent of the box length.

$$\Delta p \Delta x = m \Delta v \Delta x \geq \frac{\hbar}{2} \text{ or } \Delta v \geq \frac{\hbar}{2m\Delta x} = \frac{\hbar}{2mL}$$

$$E = \frac{1}{2}mv^2 = \frac{h^2}{8mL^2} \text{ or } v = \sqrt{\frac{h^2}{4m^2L^2}} = \frac{h}{2mL}$$

$$\frac{\Delta v}{v} = \frac{\hbar / 2mL}{h / 2mL} = \frac{1}{2\pi}$$

43. (a) $\Delta E \Delta t \geq \frac{\hbar}{2}$ so $\Delta E \geq \frac{\hbar}{2\Delta t} = \frac{6.582\times10^{-16} \text{ eV}\cdot\text{s}}{2\left(1.0\times10^{-13}\text{s}\right)} = 3.29\times10^{-3} \text{ eV}$

(b) Using the photon relation $E = \frac{hc}{\lambda}$ and taking a derivative $dE = -\frac{hc}{\lambda^2}d\lambda = -\frac{E^2}{hc}d\lambda$

Then letting $\Delta\lambda = d\lambda$ and $\Delta E = dE$ we have

$$|\Delta\lambda| = hc\frac{\Delta E}{E^2} = (1240 \text{ eV}\cdot\text{nm})\frac{3.29\times10^{-3} \text{ eV}}{(4.7 \text{ eV})^2} = 0.185 \text{nm}.$$

44. The wavelength of the electrons should be 0.14 nm or less. For this wavelength

$$p = \frac{h}{\lambda} = \frac{hc}{\lambda c} = \frac{1240 \text{ eV}\cdot\text{nm}}{(0.14 \text{ nm})c} = 8.86 \text{ keV}/c;$$

$$K = E - E_0 = \sqrt{p^2c^2 + E_0^2} - E_0 = \sqrt{(8.86 \text{ keV})^2 + (511 \text{ keV})^2} - 511 \text{ keV} = 77 \text{ eV}.$$

47. The proof is completed in Example 6.11. With $\omega = \sqrt{k/m}$ we have a minimum energy

$$E = \frac{\hbar\omega}{2} = \frac{\hbar}{2}\sqrt{\frac{k}{m}} = \frac{1.055\times10^{-34} \text{ J}\cdot\text{s}}{2}\sqrt{\frac{8.2 \text{ N/m}}{0.028 \text{ kg}}} = 9.03\times10^{-34} \text{J}.$$

57. For such highly relativistic electrons $p \approx E/c$, so the wavelength is

$$\lambda = \frac{h}{p} \approx \frac{h}{E/c} = \frac{hc}{E} = \frac{1240 \text{ eV}\cdot\text{nm}}{6.0\times10^9 \text{ eV}} = 2.1\times10^{-7} \text{ nm} = 2.1\times10^{-16} \text{ m}.$$

Typically only gamma ray photons have such high energies.

63. $\Delta t = \frac{d}{c} = \frac{1.2\times10^{-15} \text{ m}}{3.0\times10^8 \text{ m/s}} = 4.0\times10^{-24}$ s and $\Delta E \geq \frac{\hbar}{2\Delta t} = \frac{6.5821\times10^{-16} \text{ eV}\cdot\text{s}}{2\left(4.0\times10^{-24} \text{ s}\right)} = 82 \text{ MeV}.$

This "lower bound" estimate of the rest mass is $\Delta E / c^2$ which is within a factor of two of the rest energy.

© 2013 Cengage Learning. All Rights Reserved. May not be scanned, copied or duplicated, or posted to a publicly accessible website, in whole or in part.

66. Assume that the given angle corresponds to the first order reflection. We have:
$\lambda = 2d\sin\theta = 2(0.156\text{ nm})\sin 26^{\circ} = 0.1368\text{ nm}$. Next we find the energy:

$$p = \frac{h}{\lambda} = \frac{hc}{\lambda c} = \frac{1240\text{ eV}\cdot\text{nm}}{(0.1368\text{ nm})c} = 9.06\times10^{3}\text{eV}/c.$$

$$K = E - E_0 = \sqrt{p^2c^2 + E_0^2} - E_0 = \sqrt{(9.06\text{keV})^2 + (939.57\text{MeV})^2} - 939.57\text{ MeV}$$

$$= 4.36\times10^{-2}\text{eV}.$$

The Oak Ridge Electron Linear Accelerator Pulsed Neutron Source (ORELA) produces intense, nanosecond bursts of neutrons, each burst containing neutrons with energies from 10^{-3} to 10^{8} eV.

68. (a) Starting from Equation (5.45), $\Delta E\Delta t \geq \frac{\hbar}{2}$ and substituting, we have $\left(\frac{\Gamma}{2}\right)\tau \geq \frac{\hbar}{2}$.

Therefore $\Gamma\tau \geq \hbar$.

(b) We can find the minimum value for Γ from the equation above. Using the data for the neutron, for example, $\Gamma_{\text{neutron}} = \dfrac{6.5821\times10^{-16}\text{eV}\cdot\text{s}}{887\text{s}} = 7.42\times10^{-19}\text{eV}$. The other values follow in a similar fashion. $\Gamma_{\text{pion}} = 2.53\times10^{-8}\text{eV}$ and $\Gamma_{\text{upsilon}} = 65.8\text{keV}$.

© 2013 Cengage Learning. All Rights Reserved. May not be scanned, copied or duplicated, or posted to a publicly accessible website, in whole or in part.

Chapter 6

4. $\Psi^*\Psi = A^2 \exp[-i(kx-\omega t)+i(kx-\omega t)] = A^2$. The condition for normalization becomes

$$\int_0^a \Psi^*\Psi dx = A^2 \int_0^a dx = A^2 a = 1 \text{ so } A = \frac{1}{\sqrt{a}} \text{ and } \Psi = \frac{1}{\sqrt{a}}\exp[i(kx-\omega t)].$$

5. $\Psi^*\Psi = A^2 r^2 \exp\left(\frac{-2r}{\alpha}\right)$. The condition for normalization becomes

$$\int_0^\infty \Psi^*\Psi dr = A^2 \int_0^\infty r^2 \exp\left(\frac{-2r}{\alpha}\right) dr = A^2\left[\frac{2}{(2/\alpha)^3}\right] = \frac{A^2\alpha^3}{4} = 1.$$

Therefore $A = \sqrt{\frac{4}{\alpha^3}} = 2\alpha^{-3/2}$.

10. Using the Euler relations between exponential and trig functions, we find $\psi = A\left(e^{ix}+e^{-ix}\right) = 2A\cos(x)$.

Normalization: $\int_{-\pi}^{\pi} \psi^*\psi\, dx = 4A^2 \int_{-\pi}^{\pi} \cos^2(x) dx = 4A^2\pi = 1$. Thus $A = \frac{1}{2\sqrt{\pi}}$.

(a) The probability of being in the interval $[0, \pi/8]$ is

$$P = \int_0^{\pi/8} \psi^*\psi\, dx = \frac{1}{\pi}\int_0^{\pi/8} \cos^2(x) dx = \frac{1}{\pi}\left(\frac{x}{2}+\frac{1}{4}\sin(2x)\right)\Big|_0^{\pi/8}$$

$$= \frac{1}{16} + \frac{1}{4\pi\sqrt{2}} = 0.119.$$

(b) The probability of being in the interval $[0, \pi/4]$ is

$$P = \int_0^{\pi/4} \psi^*\psi\, dx = \frac{1}{\pi}\int_0^{\pi/4} \cos^2(x) dx = \frac{1}{\pi}\left(\frac{x}{2}+\frac{1}{4}\sin(2x)\right)\Big|_0^{\pi/4}$$

$$= \frac{1}{8} + \frac{1}{4\pi} = 0.205.$$

13. The wave function for the nth level is $\psi_n(x) = \sqrt{\frac{2}{L}}\sin\left(\frac{n\pi x}{L}\right)$ so the average value of the square of the wave function is

$$\left\langle \psi_n^2(x) \right\rangle = \frac{\int_0^L \psi^*\psi\, dx}{\int_0^L dx} = \frac{1}{L}\int_0^L \psi^*\psi\, dx = \frac{2}{L^2}\int_0^L \sin^2\left(\frac{n\pi x}{L}\right) dx = \frac{2}{L^2}\frac{L}{2} = \frac{1}{L}.$$

This result for the average value of the square of the wave function is independent of n and is the same as the classical probability. The classical probability is uniform throughout the box, but this is not so in the quantum mechanical case, which is $\frac{2}{L}\sin^2(k_n x)$.

© 2013 Cengage Learning. All Rights Reserved. May not be scanned, copied or duplicated, or posted to a publicly accessible website, in whole or in part.

15. (a) Starting with Equation (6.35) and using the electron mass and the length given, we have:

$$E_n = n^2\frac{\pi^2\hbar^2}{2mL^2} = n^2\frac{\pi^2(\hbar c)^2}{2(mc^2)L^2} = n^2\frac{\pi^2(197.3\text{ eV}\cdot\text{nm})^2}{2(5.11\times10^5\text{eV})(2000\text{ nm})^2}$$
$$= n^2(9.40\times10^{-8}\text{eV}).$$

Then the three lowest energy levels are: $E_1 = 9.40\times10^{-8}\text{eV}$; $E_2 = 1.88\times10^{-7}\text{eV}$; and $E_1 = 2.82\times10^{-7}\text{eV}$.

(b) Average kinetic energy equals $\frac{3}{2}kT = \frac{3}{2}\left[\left(1.381\times10^{-23}\text{ J/K}\right)\left(\frac{1\text{ eV}}{1.602\times10^{-19}\text{ J}}\right)\right](13\text{K})$

which equals $1.68\times10^{-3}\text{eV}$. Substitute this value into the equation in (a) for E_n and solve for n: $n = 134$.

19. $E_1 = \frac{h^2}{8mL^2} = \frac{h^2c^2}{8mc^2L^2} = \frac{(1240\text{ eV}\cdot\text{nm})^2}{8(511\times10^3\text{ eV})(10^{-5}\text{ nm})^2} = 3.76\text{ GeV}.$

Then $E_2 = 4E_1 = 15.05$ GeV and $\Delta E = E_2 - E_1 = 11.3$ GeV. This is much larger than any energy observed in nuclear process; the electron is not in the nucleus.

21. As in previous problems the ground state energy is

$$E_1 = \frac{h^2}{8mL^2} = \frac{h^2c^2}{8mc^2L^2} = \frac{(1240\text{ eV}\cdot\text{nm})^2}{8(511\times10^3\text{ eV})(0.70\text{ nm})^2} = 0.7676\text{ eV}.$$

The other energy levels are:

$E_n = n^2E_1$: $E_2 = 4E_1 = 3.07\text{eV}$; $E_3 = 9E_1 = 6.91\text{eV}$; $E_4 = 16E_1 = 12.28$ eV.

The allowed photon energies are:

$E_4 - E_3 = 5.37\text{eV}$ $E_4 - E_2 = 9.21\text{eV}$ $E_4 - E_1 = 11.5$ eV

$E_3 - E_2 = 3.84\text{eV}$ $E_3 - E_1 = 6.14\text{eV}$ $E_2 - E_1 = 2.30$ eV

24. Using the same notation as the text, from the boundary condition $\psi_{\text{I}}(x=0) = \psi_{\text{II}}(x=0)$ we have $Ae^0 = Ce^0 + De^0$ or $A = C + D$. From the condition $\psi_{\text{I}}'(x=0) = \psi_{\text{II}}'(x=0)$ we have $\alpha A = ikC - ikD$. Solving this last expression for A and combining with the first boundary condition gives $C + D = \frac{ik}{\alpha}C - \frac{ik}{\alpha}D$ or after rearranging $\frac{C}{D} = \frac{ik+\alpha}{ik-\alpha}$.

28. We must normalize by evaluating the triple integral of $\psi^*\psi$: $\iiint \psi^*\psi\, dxdydz = 1$ with $\psi(x,y,z)$ given by Equation (6.47) in the text. We can evaluate the iterated triple integral

© 2013 Cengage Learning. All Rights Reserved. May not be scanned, copied or duplicated, or posted to a publicly accessible website, in whole or in part.

$A^2\int_0^L \sin^2\left(\frac{\pi x}{L}\right)dx\int_0^L \sin^2\left(\frac{\pi y}{L}\right)dy\int_0^L \sin^2\left(\frac{\pi z}{L}\right)dz = A^2\left(\frac{L}{2}\right)^3 = 1$. Solving for A we find

$$A = \left(\frac{2}{L}\right)^{3/2}.$$

33. $\Delta E_n = E_{n+1} - E_n = \left(n+1+\frac{1}{2}\right)\hbar\omega - \left(n+\frac{1}{2}\right)\hbar\omega = \hbar\omega$ for all n. This is true for all n, and there is no restriction on the number of levels.

38. Taking the derivatives for the Schrödinger equation:

$$\frac{d\psi}{dx} = Ae^{-\alpha x^2/2} - A\alpha x^2 e^{-\alpha x^2/2};\ \frac{d^2\psi}{dx^2} = -3A\alpha x e^{-\alpha x^2/2} + A\alpha^2 x^3 e^{-\alpha x^2/2} = \left(\alpha^2x^2 - 3\alpha\right)\psi.$$

Combining Equation (6.56) with this, we see $\frac{d^2\psi}{dx^2} = \left(\alpha^2x^2 - \beta\right)\psi = \left(\alpha^2x^2 - 3\alpha\right)\psi$. Thus we see that $\beta = 3\alpha$ or $\frac{2mE}{\hbar^2} = 3\sqrt{\frac{mk}{\hbar^2}}$. Solving for the energy, we find

$$E = \frac{3}{2}\hbar\sqrt{\frac{k}{m}} = \frac{3}{2}\hbar\omega.$$

42. In each case $\kappa L \gg 1$ so we can use $T = 16\frac{E}{V_0}\left(1-\frac{E}{V_0}\right)e^{-2\kappa L}$ where

$$\kappa = \frac{\sqrt{2mc^2(V_0 - E)}}{\hbar c} = \frac{\left(2\left(3727\times10^6\ \text{eV}\right)\left(10\times10^6\ \text{eV}\right)\right)^{1/2}}{197.4\ \text{eV}\cdot\text{nm}} = 1.38\times10^{15}\ \text{m}^{-1}.$$

(a) With $L = 1.3\times10^{-14}$ m,

$$T_a = 16\frac{5\ \text{MeV}}{15\ \text{MeV}}\left(1-\frac{5\ \text{MeV}}{15\ \text{MeV}}\right)e^{-2\left(1.38\times10^{15}\ \text{m}^{-1}\right)\left(1.3\times10^{-14}\ \text{m}\right)} = 9.3\times10^{-16}.$$

(b) With $V_0 = 30$ MeV,

$$\kappa = \frac{\sqrt{2mc^2(V_0 - E)}}{\hbar c} = \frac{\left(2\left(3727\times10^6\ \text{eV}\right)\left(25\times10^6\ \text{eV}\right)\right)^{1/2}}{197.4\ \text{eV}\cdot\text{nm}} = 2.19\times10^{15}\ \text{m}^{-1}.$$

$$T_b = 16\frac{5\ \text{MeV}}{30\ \text{MeV}}\left(1-\frac{5\ \text{MeV}}{30\ \text{MeV}}\right)e^{-2\left(2.19\times10^{15}\ \text{m}^{-1}\right)\left(1.3\times10^{-14}\ \text{m}\right)} = 4.2\times10^{-25}.$$

(c) With $V_0 = 15$ MeV we return to the original value of κ, but now $L = 2.6\times10^{-14}$ m and

$$T_c = 16\frac{5\ \text{MeV}}{15\ \text{MeV}}\left(1-\frac{5\ \text{MeV}}{15\ \text{MeV}}\right)e^{-2\left(1.38\times10^{15}\ \text{m}^{-1}\right)\left(2.6\times10^{-14}\ \text{m}\right)} = 2.4\times10^{-31}.$$

By comparison $T_a > T_b > T_c$.

© 2013 Cengage Learning. All Rights Reserved. May not be scanned, copied or duplicated, or posted to a publicly accessible website, in whole or in part.

51. As in the text we find $E = \frac{\hbar^2\pi^2}{2m}\left(\frac{n_1^2}{L_1^2}+\frac{n_2^2}{L_2^2}+\frac{n_3^2}{L_3^2}\right)$ and substituting the given values of L we find $E = \frac{\hbar^2\pi^2}{2mL^2}\left(n_1^2 + 2n_2^2 + \frac{n_3^2}{4}\right)$. Letting $E_0 = \hbar^2\pi^2 / 2mL^2$ we have:

$$E_1 = E_0\left(1+2+\frac{1}{4}\right) = \frac{13}{4}E_0;\ E_2 = E_0\left(1+2+\frac{2^2}{4}\right) = 4E_0;\ E_3 = E_0\left(1+2+\frac{3^2}{4}\right) = \frac{21}{4}E_0;$$

$$E_4 = E_0\left(2^2+2+\frac{1}{4}\right) = \frac{25}{4}E_0;\ E_5 = E_0\left(1+2+\frac{4^2}{4}\right) = E_0\left(2^2+2+\frac{2^2}{4}\right) = 7E_0.$$

Of those listed, only E_5 is degenerate.

53. (a) In general inside the box we have a superposition of sine and cosine functions, but only the sine function satisfies the boundary condition $\psi(0) = 0$, and thus $\psi = A\sin(kx)$. With $V = 0$ inside the well, $E = \frac{p^2}{2m} = \frac{\hbar^2k^2}{2m}$ or $k = \frac{\sqrt{2mE}}{\hbar}$. Outside the well the decaying exponential is required as explained in section 6.4 of the text, with $E = \frac{\hbar^2k^2}{2m} + V_0$ which reduces to $\kappa = ik = \frac{\sqrt{2m(V_0 - E)}}{\hbar}$.

(b) Equating the wave functions and first derivatives at $x = L$:

$$A\sin(kL) = Be^{-\kappa L} \text{ and } kA\cos(kL) = -\kappa Be^{-\kappa L}.$$

Dividing the first equation by the second gives $\frac{\tan(kL)}{k} = -\frac{1}{\kappa}$ or $\kappa\tan(kL) = -k$.

57. For the one-dimensional well [Equation (6.35)],

$$E_1 = \frac{\pi^2\hbar^2}{2mL^2} = \frac{\pi^2(\hbar c)^2}{2mc^2L^2} = \frac{\pi^2(197.3\text{ eV}\cdot\text{nm})^2}{2(5.11\times10^5\text{ eV})\left[2(5.292\times10^{-2}\text{ nm})\right]^2} = 33.6\text{ eV}.$$

This is the same order of magnitude but larger than the electron's ground-state kinetic energy. The one-dimensional model is a fair but inaccurate representation of the atom.

65. The solution is identical to the presentation in the text for the three-dimensional box but without the z dimension. We assume a trial function for the form $\psi(x, y) = A\sin(k_1x)\sin(k_2y)$. Assuming that one corner is at the origin, applying the boundary conditions leads to $k_1 = \frac{n_x\pi}{L}$ and $k_2 = \frac{n_y\pi}{L}$ and substituting into the Schrödinger equation leads to $E = \frac{\pi^2\hbar^2}{2mL^2}\left(n_x^2 + n_y^2\right)$. To normalize, solve the iterated double integral:

$$\int_0^L\int_0^L \psi^*\psi\,dxdy = A^2\int_0^L\int_0^L \sin^2\left(\frac{n_x\pi x}{L}\right)\sin^2\left(\frac{n_y\pi y}{L}\right)dxdy = A^2\left(\frac{L}{2}\right)\left(\frac{L}{2}\right) = 1$$

© 2013 Cengage Learning. All Rights Reserved. May not be scanned, copied or duplicated, or posted to a publicly accessible website, in whole or in part.

so $A=\frac{2}{L}$. Now to find the energy levels use the energy equation with different values of the quantum numbers. Letting $E_0=\frac{\pi^2\hbar^2}{2mL^2}$ we have: $E_1=E_0\left(1^2+1^2\right)=2E_0$ with $n_1=1, n_2=1$;

$E_2=E_0\left(2^2+1^2\right)=5E_0$ with $n_1=2, n_2=1$ or vice versa;

$E_3=E_0\left(2^2+2^2\right)=8E_0$ with $n_1=2, n_2=2$;

$E_4=E_0\left(3^2+1^2\right)=10E_0$ with $n_1=3, n_2=1$ or vice versa;

$E_5=E_0\left(3^2+2^2\right)=13E_0$ with $n_1=3, n_2=2$ or vice versa;

$E_6=E_0\left(4^2+1^2\right)=17E_0$ with $n_1=4, n_2=1$ or vice versa.

68. (a) Note that this is an approximate procedure for one-dimensional problems with a gradually varying potential, $V(x)$. We begin with Equation (6.62b) which was derived for a scenario where $E>V$ but with V constant. We found $k=\frac{\sqrt{2m(E-V_0)}}{\hbar}$ for a constant V_0. Because our potential varies slowly, we approximate the wave number by $k=\frac{\sqrt{2m[E-V(x)]}}{\hbar}$. We also know that $p=\hbar k$ and $\lambda=\frac{h}{p}$. Combining all of the above, we find a position-dependent wavelength $\lambda(x)=\frac{h}{\sqrt{2m[E-V(x)]}}$.

(b) If we neglect barrier penetration, then the wave function must be zero at the turning points. From the particle in a box example, we know that the number of wavelengths that fit between the turning points is $\frac{1}{2}$, or 1, or $\frac{3}{2}$, etc., which equals the distance divided by the wavelength. By analogy, the number of wavelengths that can fit inside our potential well with a slowly varying wavelength is $\int\frac{dx}{\lambda(x)}=\frac{n}{2}$ where n is an integer. Substituting from above and rearranging, we have $2\int\sqrt{2m[E-V(x)]}\,dx=nh$ where n is an integer.

© 2013 Cengage Learning. All Rights Reserved. May not be scanned, copied or duplicated, or posted to a publicly accessible website, in whole or in part.

Chapter 7

1. Starting with Equation (7.7), let the electron move in a circle of radius a in the xy plane, so $\sin\theta = \sin(\pi/2) = 1$. With both r and θ constant, R and f are also constant. Let $R = f = 1$. Then $g = \psi$ and the derivatives of R and f are zero. With this Equation (7.7) reduces to $-\frac{2\mu}{\hbar^2}a^2(E-V) = \frac{1}{\psi}\frac{d^2\psi}{d\phi^2}$. In uniform circular motion with an inverse-square force, we know from the planetary model that $E = V/2$, and $E - V = \frac{V}{2} - V = -\frac{V}{2} = |E|$.

   Thus $-\frac{2\mu}{\hbar^2}a^2|E| = \frac{1}{\psi}\frac{d^2\psi}{d\phi^2}$ or $\frac{1}{a^2}\frac{d^2\psi}{d\phi^2} + \frac{2\mu}{\hbar^2}|E| = 0$.

5. Letting the constants in the front of R be called A we have $R = A\left(2 - \frac{r}{a_0}\right)e^{-r/2a_0}$,

   $\frac{dR}{dr} = A\left(-\frac{2}{a_0} + \frac{r}{2a_0^2}\right)e^{-r/2a_0}$ and $\frac{d^2R}{dr^2} = A\left(\frac{3}{2a_0^2} - \frac{r}{4a_0^3}\right)e^{-r/2a_0}$.

   Substituting these into Equation (7.13) we have

   $$\left(-\frac{1}{4a_0^3} - \frac{2\mu E}{a_0\hbar^2}\right)r + \left(\frac{5}{2a_0^2} + \frac{4\mu E}{\hbar^2} - \frac{2\mu e^2}{4\pi\epsilon_0 a_0\hbar^2}\right) + \left(-\frac{4}{a_0} + \frac{4\mu e^2}{4\pi\epsilon_0\hbar^2}\right)\frac{1}{r} = 0.$$

   To satisfy the equation, each of the expressions in parentheses must equal zero. From the $1/r$ term we find $a_0 = \frac{4\pi\epsilon_0\hbar^2}{\mu e^2}$ which is correct. From the r term we obtain

   $E = -\frac{\hbar^2}{8\mu a_0^2} = -\frac{E_0}{4}$ which is consistent with the Bohr result. The other expression in parentheses also leads directly to $E = -E_0/4$, so the solution is verified.

8. The wave function given is $\psi_{100}(r,\theta,\phi) = Ae^{-r/a_0}$ so $\psi^*\psi$ is given by $\psi_{100}{}^*\psi_{100} = A^2e^{-2r/a_0}$. To normalize the wave function, compute the triple integral over all space $\iiint \psi^*\psi\, dV = A^2\int_0^{2\pi}\int_0^{\pi}\int_0^{\infty} r^2\sin\theta e^{-2r/a_0}\,dr\,d\theta\,d\phi$. The ϕ integral yields 2π, and the θ integral yields 2. This leaves $\iiint \psi^*\psi\, dV = 4\pi A^2\int_0^{\infty} r^2 e^{-2r/a_0}\,dr = 4\pi A^2\frac{2}{(2/a_0)^3} = \pi a_0^3 A^2$

   This integral must equal 1 due to normalization which leads to $\pi a_0^3 A^2 = 1$ so $A = \frac{1}{\sqrt{\pi a_0^3}}$.

© 2013 Cengage Learning. All Rights Reserved. May not be scanned, copied or duplicated, or posted to a publicly accessible website, in whole or in part.

10. $n = 3$ and $\ell = 1$, so $m_\ell = 0$ or ± 1. Thus $L_z = 0$ or $\pm\hbar$. $L = \sqrt{\ell(\ell+1)}\,\hbar = \sqrt{2}\,\hbar$.

L_y and L_x are unrestricted except for the constraint $L_x^2 + L_y^2 = L^2 - L_z^2$.

15. $\cos\theta = \dfrac{L_z}{L} = \dfrac{m_\ell}{\sqrt{\ell(\ell+1)}}$. For this extreme case we could have $\ell = m_\ell$ so

$$\cos\left(10^\circ\right) = \frac{\ell}{\sqrt{\ell(\ell+1)}} \text{ or } \cos^2\left(10^\circ\right) = \frac{\ell^2}{\ell(\ell+1)} = \frac{\ell^2}{\ell^2+\ell}.$$

Rearranging we find $\ell = \left(\dfrac{1}{\cos^2\left(10^\circ\right)} - 1\right)^{-1} = 32.16.$

We have to round up in order to get within 10°, so $\ell = 33$.

17. $\psi_{21-1} = R_{21}Y_{1-1} = \dfrac{1}{8\sqrt{\pi}\,a_0^{3/2}}\left(\dfrac{r}{a_0}\right)e^{-r/2a_0}\sin\theta e^{-i\phi}$;

$$\psi_{210} = R_{21}Y_{10} = \frac{1}{4\sqrt{2\pi}\,a_0^{3/2}}\left(\frac{r}{a_0}\right)e^{-r/2a_0}\cos\theta;$$

$$\psi_{32-1} = R_{32}Y_{2-1} = \frac{1}{81\sqrt{\pi}\,a_0^{3/2}}\left(\frac{r^2}{a_0^2}\right)e^{-r/3a_0}\sin\theta\cos\theta e^{-i\phi}$$

21. Differentiating $E = \dfrac{hc}{\lambda}$ we find: $dE = -\dfrac{hc}{\lambda^2}d\lambda$ or $|\Delta E| = \dfrac{hc}{\lambda^2}|\Delta\lambda|$. In the normal Zeeman effect, between adjacent m_ℓ states $|\Delta E| = \mu_B B$ so $\mu_B B = \left(\dfrac{hc}{\lambda_0^2}\right)|\Delta\lambda|$ or $\Delta\lambda = \dfrac{\lambda_0^2 \mu_B B}{hc}$.

24. From Problem 21, $\Delta\lambda = \dfrac{\lambda_0^2\mu_B B}{hc}$ so the magnetic field is

$$B = \frac{hc\Delta\lambda}{\lambda_0^2\mu_B} = \frac{(1240 \text{ eV}\cdot\text{nm})(0.04 \text{ nm})}{(656.5 \text{ nm})^2\left(5.788\times10^{-5} \text{ eV/T}\right)} = 1.99 \text{ T}.$$

28. As shown in Figure 7.9 the electron spin vector cannot point in the direction of B, because its magnitude is $S = \sqrt{s(s+1)} = \sqrt{3/4}\,\hbar$ and its z-component is $S_z = m_s\hbar = \hbar/2$. If the z-component of a vector is less than the vector's magnitude, the vector does not lie along the z-axis.

© 2013 Cengage Learning. All Rights Reserved. May not be scanned, copied or duplicated, or posted to a publicly accessible website, in whole or in part.

35. We must find the maxima and minima of the function

$$P(r) = r^2 |R(r)|^2 = A^2 e^{-r/a_0} \left(2 - \frac{r}{a_0}\right)^2 r^2 = A^2 \left(4r^2 - \frac{4r^3}{a_0} + \frac{r^4}{a_0^2}\right) e^{-r/a_0}.$$

To find the extremes set $\frac{dP}{dr} = 0$:

$$0 = -\frac{1}{a_0}\left(4r^2 - \frac{4r^3}{a_0} + \frac{r^4}{a_0^2}\right) e^{-r/a_0} + \left(8r - \frac{12r^2}{a_0} + \frac{4r^3}{a_0^2}\right) e^{-r/a_0}$$

which simplifies to $0 = -\frac{r^3}{a_0^3} + \frac{8r^2}{a_0^2} - \frac{16r}{a_0} + 8.$

Letting $x = \frac{r}{a_0}$ the equation above can be factored into $(x-2)(x^2 - 6x + 4) = 0$. From the first factor we obtain $x = 2$ (or $r = 2a_0$), which from Figure 7.12 we can see is a minimum. The second parenthesis gives a quadratic equation with solutions $x = 3 \pm \sqrt{5}$, so $r = (3 \pm \sqrt{5})a_0$. These are both maxima.

44. $R = \frac{e^2}{4\pi\epsilon_0 mc^2} = \frac{1.44 \times 10^{-9}\ \text{eV} \cdot \text{m}}{511 \times 10^3\ \text{eV}} = 2.82 \times 10^{-15}\ \text{m}.$

From the angular momentum equation

$$v = \frac{3\hbar}{4mR} = \frac{3\hbar c}{4mc^2 R} c = \frac{3(197.33\ \text{eV} \cdot \text{nm})}{4(511 \times 10^3\ \text{eV})(2.82 \times 10^{-6}\ \text{nm})} c = 103c.$$

A speed of $103c$ is prohibited by the postulates of relativity.

49. The ground state energy can be obtained using the standard Rydberg formula with the reduced mass, μ, of the muonic atom: $E_0 = \frac{e^2}{8\pi\epsilon_0 a_0} = \frac{\mu e^4}{2(4\pi\epsilon_0)^2 \hbar^2}$.

Computing the reduced mass:

$$\mu = \frac{m_p m_\mu}{m_p + m_\mu} = \frac{1}{c^2} \frac{(938.27\ \text{MeV})(105.66\ \text{MeV})}{(938.27\ \text{MeV} + 105.66\text{MeV})} = 94.966\ \text{MeV}/c^2$$

Thus

$$E_0 = \frac{\mu e^4}{2(4\pi\epsilon_0)^2 \hbar^2} = \left(\frac{e^2}{4\pi\epsilon_0}\right)^2 \frac{\mu c^2}{2(\hbar^2 c^2)} = \frac{(1.44\ \text{eV} \cdot \text{nm})^2 (94.966 \times 10^6\ \text{eV})}{2(197.33\ \text{eV} \cdot \text{nm})^2} = 2.53\ \text{keV}.$$

© 2013 Cengage Learning. All Rights Reserved. May not be scanned, copied or duplicated, or posted to a publicly accessible website, in whole or in part.

50. The interaction between the magnetic moment of the proton and magnetic moment of the electron causes hyperfine splitting. The transition between the two states causes emission of a photon with energy of $E = \dfrac{hc}{\lambda} = \dfrac{(1240\,\text{eV}\cdot\text{nm})}{21\times10^7\ \text{nm}} = 5.9\times10^{-6}\,\text{eV}$.

From the uncertainty principle, we know $\Delta E \Delta t \geq \dfrac{\hbar}{2}$.

With a lifetime of $\Delta t = 1\times10^7\ \text{y}$ then

$$\Delta E \geq \frac{6.5821\times10^{-16}\,\text{eV}}{2\left(1\times10^7\,\text{y}\right)\left(3.16\times10^7\ \text{s/y}\right)} = 1.041\times10^{-30}\,\text{eV}.$$

© 2013 Cengage Learning. All Rights Reserved. May not be scanned, copied or duplicated, or posted to a publicly accessible website, in whole or in part.

Chapter 8

1. The first two electrons are in the $1s$ subshell and have $\ell = 0$, with $m_s = \pm 1/2$. The third electron is in the $2s$ subshell and has $\ell = 0$, with either $m_s = 1/2$ or $-1/2$. With four particles there are six possible interactions: the nucleus with electrons 1, 2, 3; electron 1 with electron 2; electron 1 with electron 3; or electron 2 with electron 3. In each case it is possible to have a Coulomb interaction and a magnetic moment interaction.

7. From Figure 8.4 we see that the radius of Na is about 0.16 nm. We know that for single-electron atoms $E = -\dfrac{Ze^2}{8\pi\epsilon_0 r}$. Therefore

$$Ze = -\frac{8\pi\epsilon_0 rE}{e^2}e = -2\frac{4\pi\epsilon_0}{e^2}rEe = -\frac{2(0.16\text{ nm})(-5.14\text{ eV})}{1.44\text{ eV}\cdot\text{nm}}e = 1.14e.$$

10. Ag: $[\text{Kr}]4d^{10}5s^1$, Hf: $[\text{Xe}]4f^{14}5d^2 6s^2$, Sb: $[\text{Kr}]4d^{10}5s^2 5p^3$ where the bracket represents a closed inner shell. For example, [Kr] represents $1s^2 2s^2 2p^6 3s^2 3p^6 4s^2 3d^{10} 4p^6$.

15. In the $4d$ state $\ell = 2$ and $s = 1/2$, so $j = 5/2$ or $3/2$. As usual $m_\ell = 0, \pm 1, \pm 2$. The value of m_j ranges from $-j$ to j so its possible values are $\pm 1/2$, $\pm 3/2$, and $\pm 5/2$. As always $m_s = \pm 1/2$. The two possible term notations are $4D_{5/2}$ and $4D_{3/2}$.

18. (a) The quantum number m_J ranges from $-J$ to J, or $-7/2$ to $+7/2$. Then $J_z = m_J\hbar = \pm\hbar/2, \pm 3\hbar/2, \pm 5\hbar/2, \pm 7\hbar/2$.
(b) The minimum angle occurs when m_J is at its maximum value, which is $+7/2$. Then

$$J_z = 7\hbar/2 \text{ and } \cos\theta = \frac{J_z}{J} = \frac{m_J\hbar}{\sqrt{J(J+1)}\,\hbar} = \frac{7/2}{\sqrt{(7/2)(9/2)}} = \frac{\sqrt{7}}{3} \text{ so } \theta = 28.1^\circ.$$

21. $\Delta E = \dfrac{hc}{\lambda_1} - \dfrac{hc}{\lambda_2} = (1240\,\text{eV}\cdot\text{nm})\left(\dfrac{1}{766.41\text{ nm}} - \dfrac{1}{769.90\text{ nm}}\right) = 7.334\times 10^{-3}\text{ eV}.$
As in Example 8.8 the internal magnetic field is

$$B = \frac{m\Delta E}{e\hbar} = \frac{(9.109\times 10^{-31}\text{ kg})(7.334\times 10^{-3}\text{ eV})}{(1.602\times 10^{-19}\text{ C})(6.582\times 10^{-16}\text{ eV}\cdot\text{s})} = 63.4\text{ T}.$$

© 2013 Cengage Learning. All Rights Reserved. May not be scanned, copied or duplicated, or posted to a publicly accessible website, in whole or in part.

24. As in Example 8.8

$$\Delta E = \frac{e\hbar B}{m} = \frac{(1.602\times10^{-19}\ \text{C})(6.582\times10^{-16}\ \text{eV}\cdot\text{s})(2.55\ \text{T})}{9.109\times10^{-31}\ \text{kg}} = 2.95\times10^{-4}\text{eV}.$$

26. The minimum angle corresponds to the maximum value of J_z and hence the maximum value of m_j, which is $m_j = j$. Then $\cos\theta = \frac{J_z}{J} = \frac{m_j\hbar}{\sqrt{j(j+1)}\,\hbar} = \frac{j}{\sqrt{j(j+1)}}$.

Solving for j we find $j = \frac{1}{\frac{1}{\cos^2\theta}-1} = 2.50 = 5/2$.

33. (a) $L = 0,\ S = J = 1/2;\ g = 1 + \frac{(1/2)(3/2)+(1/2)(3/2)}{2(1/2)(3/2)} = 1+1 = 2$

(b) $L = 1,\ S = 1/2$, and $J = 3/2;\ g = 1 + \frac{(3/2)(5/2)+(1/2)(3/2)-1(2)}{2(3/2)(5/2)} = 1+\frac{1}{3} = \frac{4}{3}$

(c) $L = 2,\ S = 1/2$, and $J = 5/2;\ g = 1 + \frac{(5/2)(7/2)+(1/2)(3/2)-2(3)}{2(5/2)(7/2)} = 1+\frac{1}{5} = \frac{6}{5}$

38. (a) Calcium has two 4*s* subshell electrons outside a closed $3p^6$ subshell. So the electronic configuration is $1s^2 2s^2 2p^6 3s^2 3p^6 4s^2$. Aluminum has one 3*p* electron outside the closed 3*s* level of magnesium, so its electronic configuration is $1s^2 2s^2 2p^6 3s^2 3p^1$.

(b) The *LS* coupling for calcium can be determined since both outer shell electrons have $\ell = 0$, then $L = 0$ and $S = 0$ since one electron will have spin up and one will have spin down. Then $J = L + S$ will also be 0. Therefore in spectroscopic notation, $n\,^{2S+1}L_J$, calcium will be $4\,^1S_0$. The 3*p* electron for aluminum will have $\ell = 1$. Therefore $L = 1$ and $S = 1/2$. Now $J = L \pm S$, so $J = 3/2$ or $J = 1/2$. The state with lower *J* has lower energy, so in spectroscopic notation, aluminum is $3\,^2P_{1/2}$.

© 2013 Cengage Learning. All Rights Reserved. May not be scanned, copied or duplicated, or posted to a publicly accessible website, in whole or in part.

Chapter 9

5. (a) $\int_c^{\infty} F(v)\,dv = 4\pi C \int_c^{\infty} v^2 \exp\left(-\frac{1}{2}\beta m v^2\right) dv$ with $T = 293$ K and $C = \left(\frac{\beta m}{2\pi}\right)^{3/2}$.

   (b) For example H_2 gas at $T = 293$ K we have

$$\frac{1}{2}\beta mc^2 = \frac{(1)(2)(938\times10^6\ \text{eV})}{2(8.62\times10^{-5}\ \text{eV/K})(293\ \text{K})} = 3.7\times10^{10}.$$

The exponential of the negative of this value, $\exp(-3.7\times10^{10})$, is almost 0.

7. (a) $\bar{v} = \frac{4}{\sqrt{2\pi}}\sqrt{\frac{kT}{m}} = \frac{4}{\sqrt{2\pi}}\sqrt{\frac{(1.381\times10^{-23}\ \text{J/K})\ (300\ \text{K})}{(1.675\times10^{-27}\ \text{kg})}} = 2510\ \text{m/s}$

$$v^* = \sqrt{\frac{2kT}{m}} = \sqrt{\frac{2(1.381\times10^{-23}\ \text{J/K})\ (300\ \text{K})}{(1.675\times10^{-27}\ \text{kg})}} = 2220\ \text{m/s}$$

   (b) $\bar{v} = \frac{4}{\sqrt{2\pi}}\sqrt{\frac{kT}{m}} = \frac{4}{\sqrt{2\pi}}\sqrt{\frac{(1.381\times10^{-23}\ \text{J/K})\ (630\ \text{K})}{(1.675\times10^{-27}\ \text{kg})}} = 3640\ \text{m/s}$

$$v^* = \sqrt{\frac{2kT}{m}} = \sqrt{\frac{2(1.381\times10^{-23}\ \text{J/K})\ (630\ \text{K})}{(1.675\times10^{-27}\ \text{kg})}} = 3220\ \text{m/s}$$

14. (a) Use Equation (9.8) for the translational kinetic energy of one atom and multiply by Avogadro's number for one mole.

$$K = N_A\left(\frac{3}{2}kT\right) = (6.022\times10^{23})\left(\frac{3}{2}(1.381\times10^{-23}\ \text{J/K})(273\text{K})\right) = 3406\ \text{J}.$$

   (b) Because the translational kinetic energy depends only on temperature and one mole of anything contains the same number, the answer is the same for argon or oxygen.

© 2013 Cengage Learning. All Rights Reserved. May not be scanned, copied or duplicated, or posted to a publicly accessible website, in whole or in part.

16. Starting with the distribution $F(E)=\dfrac{8\pi C}{\sqrt{2}m^{3/2}}E^{1/2}\exp(-\beta E)$ and setting $\dfrac{dF}{dE}=0$, we obtain

$$0=\frac{d}{dE}\left[E^{1/2}\exp(-\beta E)\right]=\frac{1}{2}E^{-1/2}\exp(-\beta E)-\beta E^{1/2}\exp(-\beta E).$$

Therefore $0=E^{-1/2}-2\beta E^{1/2}$ and solving for E gives the desired result $E^{*}=\dfrac{kT}{2}$.

17. The ratio of the numbers on the two levels is

$$\frac{n_2(E)}{n_1(E)}=\frac{8\exp(-\beta E_2)}{2\exp(-\beta E_1)}=4\exp(-\beta(E_2-E_1))=10^{-5}$$

Therefore $\exp(-\beta(E_2-E_1))=2.5\times10^{-6}$.

Taking the logarithm of each side gives:

$-\beta(E_2-E_1)=-\dfrac{E_2-E_1}{kT}=\ln(2.5\times10^{-6})=-12.90$. For atomic hydrogen

$E_2-E_1=\dfrac{3}{4}E_0=10.20\,\text{eV}$. Finally

$$T=-\frac{E_2-E_1}{k(-12.90)}=-\frac{10.20\text{ eV}}{(8.618\times10^{-5}\text{ eV/K})(-12.90)}=9175\text{ K}.$$

22. At first one might think it should be 0.5, but this is not quite true due to the asymmetric shape of the distribution. Starting with Equation (9.43) for $g(E)$ and using the fact that $F_{FD}\approx1$ in this range, we have

$N(E<E_F)=\int_0^{\bar{E}}g(E)(1)dE=\dfrac{3}{2}NE_F^{-3/2}\int_0^{\bar{E}}E^{1/2}dE=NE_F^{-3/2}\bar{E}^{3/2}$. Recall that $\bar{E}=\dfrac{3}{5}E_F$,

and we see that $N(E<E_F)=N\left(\dfrac{3}{5}\right)^{3/2}=0.465N$.

25. (a) As in Problem 23, $N/V=5.86\times10^{28}\text{ m}^{-3}$. Then

$$E_F=\frac{h^2}{8m}\left(\frac{3}{\pi}\frac{N}{V}\right)^{2/3}=\frac{(6.626\times10^{-34}\text{ J}\cdot\text{s})^2}{8(9.109\times10^{-31}\text{ kg})}\left(\frac{3}{\pi}(5.86\times10^{28}\text{m}^{-3})\right)^{2/3}$$

$=8.81\times10^{-19}\text{ J}=5.50\text{ eV}.$

(b) $u_F=\sqrt{\dfrac{2E_F}{m}}=\sqrt{\dfrac{2(8.81\times10^{-19}\text{ J})}{9.109\times10^{-31}\text{ kg}}}=1.39\times10^{6}\text{ m/s}.$

© 2013 Cengage Learning. All Rights Reserved. May not be scanned, copied or duplicated, or posted to a publicly accessible website, in whole or in part.

29. In general $E_F = \frac{1}{2}mu_F^2$ so $u_F = \sqrt{2E_F/m}$.

(a) $$u_F = \sqrt{\frac{2E_F}{m}} = \sqrt{\frac{2(4.69\text{ eV})(1.602\times10^{-19}\text{ J/eV})}{9.109\times10^{-31}\text{ kg}}} = 1.28\times10^6\text{ m/s}$$

(b) $$u_F = \sqrt{\frac{2E_F}{m}} = \sqrt{\frac{2(14.3\text{ eV})(1.602\times10^{-19}\text{ J/eV})}{9.109\times10^{-31}\text{ kg}}} = 2.24\times10^6\text{ m/s}$$

30. Beginning with Equation (9.34) consider the following cases as $T \to 0$:

$$E > E_F:\ \frac{E-E_F}{kT} \to \infty \text{ so } F_{FD} \to 0$$

$$E < E_F:\ \frac{E-E_F}{kT} \to -\infty \text{ so } F_{FD} \to 1$$

$$E = E_F:\ \frac{E-E_F}{kT} \to 0 \text{ so } F_{FD} \to \frac{1}{2}$$

37. We assume that the collection of fermions behaves like an ideal gas. Using Maxwell-Boltzmann statistics, we know that $E = \frac{3}{2}kT$ or $\beta E = 3/2$. We note that $\exp(\beta E) = \exp(3/2) = 4.4817$. Now the MB factor is $\exp(-\beta E)$ and we want this to be within 1 % of the FD factor: $F_{FD} = \dfrac{1}{\exp\left[\dfrac{(E-E_F)}{kT}\right]+1}$. So we want

$\exp[\beta(E-E_F)]+1 = (1.01)\cdot(4.4817) = 4.5265$ or $[\beta(E-E_F)] = \ln(3.5265)$.

At room temperature $(\beta)^{-1} = 2.525\times10^{-2}$ eV, so the expression above can be solved to give $(E-E_F) = 3.2\times10^{-2}$ eV. This small value is possible at room temperature.

39. (a) Use the same method as in the text for helium. The liquid density of 1200 kg/m^3 translates to a number density of about 3.6×10^{28}/m^3. Plugging this into the usual formula (9.65) gives $T > 0.87$ K.

(b) Neon is not a liquid at that temperature, so it cannot be a superfluid.

40. (a) To find N/V integrate $n(E)dE$ over the whole range of energies:

$\dfrac{N}{V} = \int_0^\infty n(E)dE = \dfrac{8\pi}{h^3c^3}\int_0^\infty \dfrac{E^2}{\exp(E/kT)-1}dE$. From integral tables we have the following:

© 2013 Cengage Learning. All Rights Reserved. May not be scanned, copied or duplicated, or posted to a publicly accessible website, in whole or in part.

$\int_0^\infty \frac{x^{n-1}}{e^{mx}-1}dx = m^{-n}\Gamma(n)\zeta(n)$. For our integral $m = \frac{1}{kT}$, $\Gamma(3) = 2! = 2$, and from numerical tables $\zeta(3) \approx 1.20$. Therefore $\frac{N}{V} = \frac{8\pi}{h^3c^3}(kT)^3(2)(1.20) = \frac{8\pi k^3T^3}{h^3c^3}(2.40)$.

(b) With $T = 500$ K:

$$\frac{N}{V} = \frac{8\pi k^3T^3}{h^3c^3}(2.40) = 8\pi(2.40)\left(\frac{(1.381\times10^{-23}\text{ J/K})(500\text{ K})}{(6.626\times10^{-34}\text{ J}\cdot\text{s})(2.998\times10^8\text{ m/s})}\right)^3$$

$$= 2.53\times10^{15}\text{ m}^{-3}.$$

(c) With $T = 5500$ K:

$$\frac{N}{V} = \frac{8\pi k^3T^3}{h^3c^3}(2.40) = 8\pi(2.40)\left(\frac{(1.381\times10^{-23}\text{ J/K})(5500\text{ K})}{(6.626\times10^{-34}\text{ J}\cdot\text{s})(2.998\times10^8\text{ m/s})}\right)^3$$

$$= 3.37\times10^{18}\text{ m}^{-3}.$$

47. The number of molecules with speed v that hit the wall per unit time is proportional to v and $F(v)$, so that the distribution $W(v)$ of the escaping molecules is by proportion

$$W(v) \sim vF(v) \sim v^3\exp\left(-\frac{1}{2}\beta mv^2\right).$$

Let the normalization constant for $W(v)$ be C', so

$$C'\int_0^\infty v^3\exp\left(-\frac{1}{2}\beta mv^2\right)dv = 1 = C'\left(\frac{1}{2}\right)\left(\frac{\beta m}{2}\right)^{-2} \text{ or } C' = \beta^2m^2/2.$$

The mean kinetic energy of the escaping molecules is

$$\bar{E} = \frac{1}{2}m\overline{v^2} = \frac{1}{2}mC'\int_0^\infty v^5\exp\left(-\frac{1}{2}\beta mv^2\right)dv = \frac{1}{2}m\left(\frac{\beta^2m^2}{2}\right)\left(\frac{\beta m}{2}\right)^{-3} = \frac{2}{\beta} = 2kT.$$

53. For the harmonic oscillator the position and velocity are $x = x_0\cos(\omega t)$ and $v = \frac{dx}{dt} = -\omega x_0\sin(\omega t)$ respectively. $V = \frac{1}{2}kx^2 = \frac{1}{2}kx_0^2\cos^2(\omega t)$;

$$K = \frac{1}{2}mv^2 = \frac{1}{2}m\omega^2x_0^2\sin^2(\omega t) = \frac{1}{2}kx_0^2\sin^2(\omega t)$$

where we have used the fact that $\omega^2m = k$. Over one cycle the average of the square of the sine or cosine function is one-half. Also the total energy is $E = \frac{1}{2}kx_0^2$. Therefore

$$\bar{K} = \bar{V} = \frac{1}{2}kx_0^2\left(\frac{1}{2}\right) = \frac{E}{2}.$$

© 2013 Cengage Learning. All Rights Reserved. May not be scanned, copied or duplicated, or posted to a publicly accessible website, in whole or in part.

57. Rearranging Equation (9.64) and with $m = 4$ u, we have

$$\frac{N}{V} \le \frac{2\pi(2.315)}{h^3}\left[2mkT\right]^{3/2}$$

$$\le \frac{2\pi(2.315)}{\left(6.626\times10^{-34}\,\text{J·s}\right)^3}\left[2(4)\left(1.6605\times10^{-27}\,\text{kg}\right)\left(1.381\times10^{-23}\,\text{J/K}\right)(293)\right]^{3/2}$$

$$\le 1.97\times10^{31}\,\text{m}^{-3}.$$

The number density of an ideal gas at STP is $\dfrac{N}{V} = \dfrac{P}{kT} = \dfrac{1.0135\times10^{5}\,\text{Pa}}{\left(1.381\times10^{-23}\,\text{J/K}\right)\left(293\text{ K}\right)}$.

Therefore $\dfrac{N}{V} = 2.50\times10^{25}\,\text{m}^{-3}$. As you would expect, the condensate has a number density nearly one million times greater than the ideal gas.

© 2013 Cengage Learning. All Rights Reserved. May not be scanned, copied or duplicated, or posted to a publicly accessible website, in whole or in part.

Chapter 10

1.

(a) For each state the energy is given by $E_{\text{rot}} = \dfrac{\hbar^2 \ell(\ell+1)}{2I}$, so the transition energy is

$$\Delta E = \frac{\hbar^2}{2I}\left[2(3)-1(2)\right] = \frac{2\hbar^2}{I} = \frac{2\left(1.055\times10^{-34}\ \text{J}\cdot\text{s}\right)^2}{10^{-46}\,\text{kg}\cdot\text{m}^2} = 2.2\times10^{-22}\ \text{J} = 1.4\times10^{-3}\ \text{eV}.$$

(b) As in part (a), $\Delta E = \dfrac{\hbar^2}{2I}\left(10(11)-9(10)\right) = \dfrac{31\hbar^2}{2I} = 1.73\times10^{-21}\ \text{J} = 1.08\times10^{-2}\,\text{eV}.$

This is still in the infrared part of the spectrum.

5. With Bohr's condition $L = n\hbar$ we find $E_{\text{rot}} = \dfrac{L^2}{2I} = \dfrac{n^2\hbar^2}{2I}$. The Bohr version and the correct version become similar for large values of quantum number n or ℓ, but they are quite different for small ℓ.

6. $\Delta E = E_1 - E_0 = E_1 = \dfrac{\hbar^2}{I} = \dfrac{hc}{\lambda}$.

(a) $I = \dfrac{\hbar^2\lambda}{hc} = \dfrac{\hbar\lambda}{2\pi c} = \dfrac{\left(1.055\times10^{-34}\ \text{J}\cdot\text{s}\right)\left(2.60\times10^{-3}\ \text{m}\right)}{2\pi\left(2.998\times10^{8}\ \text{m/s}\right)} = 1.46\times10^{-46}\ \text{kg}\cdot\text{m}^2.$

(b) The minimum energy in a vibrational transition $\Delta E = hf$. From Table 10.1 $f = 6.42\times10^{13}$ Hz, which corresponds to a photon of wavelength $\lambda = c/f = 4.67\,\mu\text{m}$. A photon of this wavelength or less is required to excite the vibrational mode, so the 2.60-mm photon is too weak.

10. (a) The distance of each H atom from the line is $d = (0.0958\text{nm})\left(\sin 52.5^\circ\right) = 7.60\times10^{-2}\,\text{nm}.$

Then $I = 2m_{\text{H}}d^2 = 2\left(1.67\times10^{-27}\ \text{kg}\right)\left(7.60\times10^{-11}\ \text{m}\right)^2 = 1.93\times10^{-47}\ \text{kg}\cdot\text{m}^2$

(b) $E_1 = \dfrac{\hbar^2}{I} = \dfrac{\left(1.055\times10^{-34}\ \text{J}\cdot\text{s}\right)^2}{1.93\times10^{-47}\ \text{kg}\cdot\text{m}^2} = 5.77\times10^{-22}\ \text{J} = 3.61\ \text{MeV}$

$E_2 = \dfrac{3\hbar^2}{I} = \dfrac{3\left(1.055\times10^{-34}\ \text{J}\cdot\text{s}\right)^2}{1.93\times10^{-47}\ \text{kg}\cdot\text{m}^2} = 1.73\times10^{-21}\ \text{J} = 10.81\ \text{MeV}$

(c) $\lambda = \dfrac{hc}{E_1} = \dfrac{\left(6.626\times10^{-34}\ \text{J}\cdot\text{s}\right)\left(2.998\times10^{8}\ \text{m/s}\right)}{5.77\times10^{-22}\ \text{J}} = 344\ \mu\text{m}$

© 2013 Cengage Learning. All Rights Reserved. May not be scanned, copied or duplicated, or posted to a publicly accessible website, in whole or in part.

15. The gap between adjacent lines is $h(\Delta f) = \hbar^2 / I$.

(a) $I = \dfrac{\hbar^2}{h(\Delta f)} = \dfrac{\left(1.055\times10^{-34}\ \text{J}\cdot\text{s}\right)^2}{\left(6.626\times10^{-34}\ \text{J}\cdot\text{s}\right)\left(7\times10^{11}\ \text{Hz}\right)} = 2.4\times10^{-47}\ \text{kg}\cdot\text{m}^2$

(b) $\omega = 2\pi f = \sqrt{\kappa/\mu}$ with $f = 8.65\times10^{13}$ Hz.

The reduced mass is (using the ^{35}Cl isotope) $\mu = \dfrac{m_1 m_2}{m_1 + m_2} = \dfrac{35}{36}\ \text{u} = 1.614\times10^{-27}\ \text{kg}$.

Solving for κ we find

$\kappa = 4\pi^2 f^2 \mu = 4\pi^2\left(8.65\times10^{13}\text{Hz}\right)^2\left(1.614\times10^{-27}\ \text{kg}\right) = 478\ \text{N/m}$

in good agreement with Table 10.1.

18. (a) Using dimensional analysis and the fact that the energy of each photon is

$hc/\lambda = 3.14\times10^{-19}\,\text{J}$, then $\dfrac{N}{t} = \dfrac{5\times10^{-3}\ \text{J/s}}{3.14\times10^{-19}\ \text{J/photon}} = 1.59\times10^{16}\,\text{photon/s}$.

(b) 0.02 mole is equivalent to $0.02N_A = 1.20\times10^{22}$ atoms. The fraction participating is

$\dfrac{1.59\times10^{16}}{1.20\times10^{22}} = 1.33\times10^{-6}$.

(c) The transitions involved have a fairly low probability, even with stimulated emission. We are saved by the large number of atoms available.

19. (a) The energy of each photon is $E_2 - E_1 = \dfrac{1.15\ \text{eV}}{\text{photon}}\cdot\dfrac{1.602\times10^{-19}\,\text{J}}{1\text{eV}} = 1.84\times10^{-19}\ \text{J/photon}$.

Because $P = E/t$, we have $P = \left(1.84\times10^{-19}\,\text{J/photon}\right)\left(5.50\times10^{18}\,\text{photons/s}\right) = 1.00\,\text{W}$.

(b) $\lambda = \dfrac{hc}{\Delta E} = \dfrac{1.986\times10^{-25}\,\text{J}\cdot\text{m}}{1.84\times10^{-19}\,\text{J}} = 1.08\times10^{-6}\,\text{m} = 1.08\ \mu\text{m}$.

25. Using dimensional analysis the number density is

$1980\ \text{kg/m}^3\ \dfrac{1\ \text{mol}}{0.07455\ \text{kg}}\ \dfrac{2\left(6.022\times10^{23}\right)}{\text{mol}} = 3.20\times10^{28}\ \text{m}^{-3}$. Therefore the distance is

$d = \left(3.20\times10^{28}\ \text{m}^{-3}\right)^{-1/3} = 3.15\times10^{-10}\ \text{m} = 0.315\ \text{nm}$.

© 2013 Cengage Learning. All Rights Reserved. May not be scanned, copied or duplicated, or posted to a publicly accessible website, in whole or in part.

26. Each charge has two unlike charges a distance r away, two like charges a distance $2r$ away, and so on: $V = -\frac{2e^2}{4\pi\epsilon_0 r}\left(1 - \frac{1}{2} + \frac{1}{3} - \frac{1}{4} + \ldots\right)$. The bracketed expression is the Taylor series expansion for ln 2, so $V = -\frac{2e^2}{4\pi\epsilon_0 r}\ln 2 = -\frac{\alpha e^2}{4\pi\epsilon_0 r}$ and we see that $\alpha = 2\ln 2$.

32. We begin with Equation (10.32): $L \equiv \frac{K}{\sigma T} = \frac{4k^2}{\pi e^2}$. Solving for the thermal conductivity K,

$$K = \frac{4\sigma k^2 T}{\pi e^2} = \frac{4\left(6.30\times10^7\,\Omega^{-1}\cdot\text{m}^{-1}\right)\left(1.38\times10^{-23}\,\text{J}\cdot\text{K}^{-1}\right)^2(293\text{K})}{\pi\left(1.60\times10^{-19}\,\text{C}\right)^2} = 175\ \text{W}\cdot\text{K}^{-1}\cdot\text{m}^{-1}.$$

As mentioned in the text, the Wiedemann-Franz Law was derived using classical expressions for the mean speed and the molar heat capacity. Quantum mechanical corrections give an additional factor of $\pi^3/12 = 2.58$. With this correction, our answer would be $451\ \text{W}\cdot\text{K}^{-1}\cdot\text{m}^{-1}$ which is much closer to the measured value.

34. (a) Substituting the results from Equations (10.24) and (10.25) into Equation (10.23), we have $\dfrac{\frac{3\sqrt{\pi}}{4}ba^{-5/2}\beta^{-3/2}}{\pi^{1/2}a^{-1/2}\beta^{-1/2}} = \frac{3}{4}ba^{-2}\beta^{-1} = \frac{3bkT}{4a^2}$.

(b) Let $\frac{3bk}{4a^2} = C_0$. Then $\langle x\rangle = C_0 T$.

$$C_0 = \frac{\Delta\langle x\rangle}{\Delta T} = \alpha x = \left(1.67\times10^{-5}\ \text{K}^{-1}\right)\left(8.47\times10^{28}\ \text{m}^{-3}\right)^{-1/3} = 3.80\times10^{-15}\ \text{m/K}$$

where we used the number density of copper from Table 9.3.

39. (a) Following the arguments in the text, regardless of the sense of rotation of the electron, as the magnetic field increases from zero, the flux upward through the loop increases. Then, as in the text, the tangential electric field is still directed clockwise. The torque does not depend on the sense of rotation either, so the torque is directed out of the page. Thus the vector $\Delta\vec{L}$ is out of the page which means that $\Delta\mu = -\frac{e}{2m}\Delta L = \frac{e^2r^2B}{4m}$, and $\Delta\vec{\mu}$ is directed *into* the page and thus opposite the direction of B as before. (If the reader is uncertain that the sense of rotation is irrelevant, recall that $\Phi_B = \int \vec{B}\cdot d\vec{a}$. You use the right-hand rule to determine the sense of $d\vec{a}$. While the direction of $d\vec{a}$ will depend on the sense of rotation, the other side of the equation for Faraday's law contains the expression $\oint \vec{E}\cdot d\vec{\ell}$. The sense of transit along the increment $d\vec{\ell}$ will also change.)

© 2013 Cengage Learning. All Rights Reserved. May not be scanned, copied or duplicated, or posted to a publicly accessible website, in whole or in part.

(b) With the field directed into the paper, now the magnetic flux increases downward through the loop. Thus Faraday's law will indicate that the electric field is tangent to the orbit but now directed counterclockwise. Thus the torque will directed into the page and thus $\Delta\vec{L}$ will be directed into the page. Finally the negative sign in $\Delta\vec{\mu} = -\frac{e}{2m}\Delta\vec{L}$ will mean that $\Delta\mu = \frac{e^2r^2B}{4m}$, with $\Delta\vec{\mu}$ directed *out of* the page.

43. The magnetic dipole moment has units of $\mathrm{A\cdot m^2}$, so M has units of $\mathrm{A\cdot m^2/m^3 = A/m}$. μ_0 has units of $\mathrm{T\cdot m/A}$ and B has units T, so $\chi = \frac{\mu_0 M}{B}$ has units $\frac{(\mathrm{T\cdot m/A})(\mathrm{A/m})}{\mathrm{T}}$ which reduces to no unit.

45. $B_c = B_c(0)\left(1-\left(\frac{T}{T_c}\right)^2\right) = 0.25B_c(0)$. Thus $(T/T_c)^2 = 0.75$ and $T = \sqrt{0.75}T_c \approx 0.87T_c$.

Similarly for a ratio of 0.50 we find $T = \sqrt{0.50}T_c \approx 0.71T_c$; for a ratio of 0.75 we find $T = \sqrt{0.25}T_c \approx 0.50T_c$.

47. Using the value given in the text just below Equation (10.43), $T_c = 4.146$ K for a mass of 203.4 u, we find $M^{0.5}T_c =$ constant $= 59.1296\ \mathrm{u^{0.5}\cdot K}$. For a mass of 201 u we find $T_c = 4.171\,\mathrm{K}$ and for mass of 204 u we find $T_c = 4.140\,\mathrm{K}$.

50. $B = \mu_0 I n = (4\pi\times10^{-7}\ \mathrm{N/A^2})(4.5\ \mathrm{A})(2500\ \mathrm{m^{-1}}) = 14.1\ \mathrm{mT}$

$$\Phi = BA = \frac{B\pi d^2}{4} = (14.1\times10^{-3}\mathrm{T})\frac{\pi(0.032\ \mathrm{m})^2}{4} = 1.13\times10^{-5}\ \mathrm{T\cdot m^2}$$

$$\frac{\Phi}{\Phi_0} = \frac{1.13\times10^{-5}\ \mathrm{T\cdot m^2}}{2.068\times10^{-15}\ \mathrm{T\cdot m^2}} = 5.5\times10^9 \text{ flux quanta.}$$

This large number shows how small the flux quantum is.

64. (a) In an RL circuit the current is $I = I_0 e^{-Rt/L}$. For small values of R we approximate the exponential with the Taylor expansion $1 - Rt/L$ Then $10^{-9} = 1 - \frac{I}{I_0} = 1 - e^{-Rt/L} \approx \frac{Rt}{L}$;

$$R \le 10^{-9}\frac{L}{t} = 10^{-9}\left(\frac{3.14\times10^{-8}\ \mathrm{H}}{2.5\ \mathrm{y}\ (3.16\times10^7\ \mathrm{s/y})}\right) = 4.0\times10^{-25}\ \Omega.$$

(b) For a 10% loss, $t = \frac{0.1}{10^{-9}}(2.5\ \mathrm{y}) = 2.5\times10^8\ \mathrm{y}$.

© 2013 Cengage Learning. All Rights Reserved. May not be scanned, copied or duplicated, or posted to a publicly accessible website, in whole or in part.

Chapter 11

4. (a) Starting from Equation (11.6) and with $A = yz$, we have

$$n = \frac{IB}{eV_{\rm H}z} = \frac{(0.10\ \text{A})(0.050\ \text{T})}{\left(1.602\times10^{-19}\ \text{C}\right)\left(11.5\times10^{-3}\ \text{V}\right)\left(1.5\times10^{-4}\ \text{m}\right)} = 1.81\times10^{22}\ \text{m}^{-3}$$

(b) Graphing B versus $V_{\rm H}$ we find a slope of approximately 4.56 T/V.

Algebraically we see that $B = \frac{nez}{I}V_{\rm H}$. Thus the slope, m is equal to $\frac{nez}{I}$. So

$$n = \frac{mI}{ez} = \frac{(4.56\ \text{T/V})(0.10\ \text{A})}{\left(1.602\times10^{-19}\ \text{C}\right)\left(1.5\times10^{-4}\ \text{m}\right)} = 1.90\times10^{22}\ \text{m}^{-3}.$$

8. (a) P is in Group V; Ge is in Group IV; so adding phosphorous will create an *n*-type semiconductor.
(b) Ga is in Group III; Ge is in Group IV; so adding gallium will create a *p*-type semiconductor.

15. In general $I = I_0\left(\exp(eV/kT) - 1\right)$ and in Example 11.4

$$I_0 = \frac{I}{\exp(eV/kT) - 1} \approx 18.16\ \mu\text{A}.$$

(a) $$I = \left(18.16\times10^{-6}\text{A}\right)\left(\exp\left(\frac{0.250\ \text{eV}}{\left(8.617\times10^{-5}\ \text{eV/K}\right)(250\ \text{K})}\right) - 1\right) = 1.99\text{A}$$

(b) $$I = \left(18.16\times10^{-6}\text{A}\right)\left(\exp\left(\frac{0.250\ \text{eV}}{\left(8.617\times10^{-5}\ \text{eV/K}\right)(300\ \text{K})}\right) - 1\right) = 0.288\ \text{A} = 288\ \text{mA}$$

(c) $$I = \left(18.16\times10^{-6}\text{A}\right)\left(\exp\left(\frac{0.250\ \text{eV}}{\left(8.617\times10^{-5}\ \text{eV/K}\right)(500\ \text{K})}\right) - 1\right) = 0.006\ \text{A} = 6.00\ \text{mA}$$

16. Because the tube is single-walled, we can find the surface area density, σ in SI units. Each atom has mass 12 u, with $\text{u} = 1.6605\times10^{-27}\,\text{kg}$.

$$\sigma = \left(2.3\times10^{19}\,\text{atoms/m}^2\right)(12\,\text{u})\left(1.6605\times10^{-27}\,\text{kg/u}\right) = 4.58\times10^{-7}\,\text{kg/m}^2.$$

(a) To determine the density of the material, we need the mass per unit volume. The mass equals the mass density σ (from above) times the area A of the cylindrical wall. If the tube's length is L with radius R, them $m = 2\pi RL\sigma$. Therefore the density is

$$\rho = \frac{m}{V} = \frac{2\pi RL\sigma}{\pi R^2 L} = \frac{2\sigma}{R} = \frac{2\left(4.58\times10^{-7}\,\text{kg/m}^2\right)}{0.7\times10^{-9}\,\text{m}} = 1300\ \text{kg/m}^3.$$

(b) The material is less dense than steel and has a greater tensile strength.

© 2013 Cengage Learning. All Rights Reserved. May not be scanned, copied or duplicated, or posted to a publicly accessible website, in whole or in part.

19. (a) Replacing ϵ_0 with $\kappa\epsilon_0$, we have for the new Bohr radius

$$a_0' = \frac{4\pi\epsilon_0\kappa\hbar^2}{me^2} = 11.7a_0 = 11.7\left(5.29\times10^{-2}\ \text{nm}\right) = 0.619\ \text{nm}.$$

(b) This value is about 2.6 times the lattice spacing. This is consistent with the fact that the electron is very weakly bound, and hence the doped silicon should have a higher electrical conductivity than pure silicon.

22. (a) $I = I_0\left(\exp\left(eV/kT\right)-1\right)$. To find the value of V for the diode, use the loop rule: $V + IR = 6\text{V}$ so $V = 6\text{V} - IR$. We are given that $I_0 = 1.05\,\mu\text{A}$ and $I = 140$ mA with $T = 293$ K so $\frac{I}{I_0} = \frac{140\times10^{-3}}{1.05\times10^{-6}} = 1.33\times10^5 = \exp\left(eV/kT\right)$. Therefore

$\ln\left(1.33\times10^5\right) = \frac{eV}{kT} = \frac{e\left(6\text{V} - IR\right)}{kT}$. Solving for R we find

$$R = \frac{6\text{V} - \left(kT\ln\left(1.33\times10^5\right)\right)/e}{I} = \frac{6\text{V} - \left(8.617\times10^{-5}\text{eV/K}\right)\left[293\text{K}\left(\ln\left(1.33\times10^5\right)\right)\right]/e}{0.140\text{A}} = 40.7\,\Omega.$$

(b) $V_R = IR = \left(0.140\ \text{A}\right)\left(40.7\ \Omega\right) = 5.70\,\text{V}.$

24. $E_g = \frac{hc}{\lambda} = \frac{1240\ \text{eV}\cdot\text{nm}}{650\ \text{nm}} = 1.91\ \text{eV}.$

26. (a) The total area is $\left(580\times10^6\right)\left(45\times10^{-9}\ \text{m}\right)^2 = 1.12\times10^{-6}\text{m}^2$ so each side is the square root of this which equals 1.08 mm.

(b) The area of each transistor is now $\left(32\times10^{-9}\ \text{m}\right)^2 = 1.024\times10^{-15}\text{m}^2$ so the number is

$N = \frac{1.12\times10^{-6}\ \text{m}^2}{1.024\times10^{-15}\text{m}^2} = 1.09\times10^9$ or nearly a factor of 2 improvement.

© 2013 Cengage Learning. All Rights Reserved. May not be scanned, copied or duplicated, or posted to a publicly accessible website, in whole or in part.

31. From the diode equation the current is $I = I_0 \exp(eV/kT) - 1$. Then the ratio is

$$\frac{I_f}{I_r} = \frac{I_0 \exp(eV_f/kT) - 1}{I_0 \exp(eV_r/kT) - 1} = \frac{\exp(eV_f/kT) - 1}{\exp(eV_r/kT) - 1}.$$ Substituting the given values we find

$$\frac{I_f}{I_r} = \frac{\exp\left(\frac{0.25\,\text{eV}}{(8.617\times10^{-5}\ \text{eV/K})(293\ \text{K})}\right) - 1}{\exp\left(\frac{-0.25\,\text{eV}}{(8.617\times10^{-5}\ \text{eV/K})(293\ \text{K})}\right) - 1} = -20000.$$

The ratio of the currents in the forward- and reverse-bias is 20,000.

32. Young's Modulus relates the elongation of an object due to an applied force. The formula is $Y = \frac{F}{A}\left(\frac{L}{\Delta L}\right)$. Rearranging this formula to solve for F and substituting, we find

$$F = YA\left(\frac{\Delta L}{L}\right) = 1050\times10^9\,\text{Pa}\left(\frac{\pi(1.6\times10^{-9}\,\text{m})^2}{4}\right)(0.01) = 2.1\times10^{-8}\,\text{N}.$$

The number of bits stored is $(4.7\times10^9)(8) = 3.76\times10^{10}\,\text{bits}$. To find the area we use $A = \pi\left(r_2^2 - r_1^2\right) = \pi\left(0.058^2 - 0.023^2\right) = 8.91\times10^{-3}\,\text{m}^2$. The number of bits stored per square meter is then $\frac{3.76\times10^{10}\,\text{bits}}{8.91\times10^{-3}\,\text{m}^2} = 4.22\times10^{12}\,\text{bits/m}^2$.

33. 25 GB $\times$ 8 bits/byte = 200 Gbits With an inner radius of 2.3 cm and outer radius 5.8 cm, the effective useful area is $\pi(0.058\ \text{m})^2 - \pi(0.023\ \text{m})^2 = 8.91\times10^{-3}\,\text{m}^2$. The average area per bit is $\frac{8.91\times10^{-3}\,\text{m}^2}{200\times10^9} = 4.46\times10^{-14}\,\text{m}^2$.

For an approximately rectangular bit, the area is length times width, so the length is area divided by width, or

$$\frac{4.46\times10^{-14}\,\text{m}^2}{3.2\times10^{-7}\,\text{m}} = 1.4\times10^{-7}\,\text{m} = 0.14\ \mu\text{m}.$$

© 2013 Cengage Learning. All Rights Reserved. May not be scanned, copied or duplicated, or posted to a publicly accessible website, in whole or in part.

Chapter 12

6. $^{14}\text{N}\,(99.63\%)$, $^{15}\text{N}\,(0.37\%)$;
   $^{50}\text{V}\,(0.250\%)$, $^{51}\text{V}\,(99.75\%)$;
   $^{84}\text{Sr}\,(0.56\%)$, $^{86}\text{Sr}\,(9.86\%)$, $^{87}\text{Sr}\,(7.00\%)$, $^{88}\text{Sr}\,(82.58\%)$

9. $$\frac{\mu_p}{\mu_e} = \frac{2.79\mu_\text{N}}{-1.00116\mu_\text{B}} = -2.786\frac{\mu_\text{N}}{\mu_\text{B}} = -2.787\frac{m_e}{m_p} = -2.786\frac{0.51100}{938.27} = -1.52\times10^{-3}$$

10. From Appendix 8 the mass of the nuclide is 55.935 u or 9.29×10^{-26} kg.
$$\rho = \frac{m}{\frac{4}{3}\pi r^3} = \frac{m}{\frac{4}{3}\pi r_0^3 A} = \frac{9.29\times10^{-26}\text{ kg}}{\frac{4}{3}\pi\left(1.2\times10^{-15}\text{ m}\right)^3(56)} = 2.29\times10^{17}\text{ kg/m}^3.$$

15. The distance equals the nuclear radius: $r = r_0 A^{1/3} = (1.2\text{ fm})(3^{1/3}) = 1.73\,\text{fm}$.
$$|F_g| = \frac{GMm}{r^2} = \frac{\left(6.673\times10^{-11}\text{ N}\cdot\text{m}^2/\text{kg}^2\right)\left(1.673\times10^{-27}\text{ kg}\right)^2}{\left(1.73\times10^{-15}\text{ m}\right)^2} = 6.24\times10^{-35}\,\text{N}$$
$$|F_e| = \frac{ke^2}{r^2} = \frac{\left(8.988\times10^{9}\text{ N}\cdot\text{m}^2/\text{C}^2\right)\left(1.602\times10^{-19}\text{ C}\right)^2}{\left(1.73\times10^{-15}\text{ m}\right)^2} = 77.1\,\text{N}$$
To compare with the strong force, we need the potential energy:
$$|V_g| = \frac{GMm}{r} = \frac{\left(6.673\times10^{-11}\text{ N}\cdot\text{m}^2/\text{kg}^2\right)\left(1.673\times10^{-27}\text{ kg}\right)^2}{\left(1.73\times10^{-15}\text{ m}\right)\left(1.602\times10^{-13}\text{ J/MeV}\right)} = 6.7\times10^{-37}\,\text{MeV}$$
$$|V_e| = \frac{ke^2}{r} = \frac{\left(8.988\times10^{9}\text{ N}\cdot\text{m}^2/\text{C}^2\right)\left(1.602\times10^{-19}\text{ C}\right)^2}{\left(1.73\times10^{-15}\text{ m}\right)\left(1.602\times10^{-13}\text{ J/MeV}\right)} = 0.83\,\text{MeV}$$
The electrostatic force is about 50 times weaker than the strong force. The gravitational force is almost 10^{38} times weaker than the strong force.

22. For ^{4}He the radius is $r = r_0 A^{1/3} = (1.2\text{ fm})(4^{1/3}) = 1.90\,\text{fm}$.
$$V = \frac{e^2}{4\pi\epsilon_0}\frac{1}{r} = \frac{1.44\times10^{-9}\text{ eV}\cdot\text{m}}{1.90\times10^{-15}\text{ m}} = 0.76\text{ MeV}$$
For ^{40}Ca:
$$\Delta E_\text{Coul} = \frac{3}{5}\frac{Z(Z-1)e^2}{4\pi\epsilon_0 R} = 0.72Z(Z-1)A^{-1/3}\text{ MeV}$$
$$= 0.72(20)(19)40^{-1/3}\,\text{MeV} = 80\,\text{MeV}.$$

© 2013 Cengage Learning. All Rights Reserved. May not be scanned, copied or duplicated, or posted to a publicly accessible website, in whole or in part.

For ^{208}Pb: $\Delta E_{\text{Coul}} = 0.72(82)(81)208^{-1/3}\ \text{MeV} = 807\,\text{MeV}$.
There is roughly a factor of ten between each of these three nuclides.

25. We begin with Equation (12.20) and substitute Equation (12.19) for the Coulomb term.

Therefore, we have: $B\left({}_Z^A\text{X}\right) = a_V A - a_A A^{2/3} - 0.72\left[Z(Z-1)\right]A^{-1/3} - a_S \dfrac{(N-Z)^2}{A} + \delta$

Evaluating this expression for ^{48}Ca, we have:

$$B\left({}_{20}^{48}\text{Ca}\right) = (14\text{MeV})48 - (13\text{MeV})48^{2/3} - 0.72\frac{[20\cdot 19]}{48^{1/3}} - (19\text{MeV})\frac{(28-20)^2}{48} + \frac{33\text{MeV}}{48^{3/4}}.$$

This gives $B\left({}_{20}^{48}\text{Ca}\right) = 401.492\,\text{MeV}$ which equals

$401.492\ \text{MeV}\left(\dfrac{c^2\cdot\text{u}}{931.5\text{MeV}}\right) = 0.43102\,\text{u}\cdot c^2$. From Equation (12.10), we can calculate the mass: $M\left({}_{20}^{48}\text{Ca}\right) = 28\cdot m_n + 20\cdot M\left({}^1\text{H}\right) - \left[B\left({}_{20}^{48}\text{Ca}\right)/c^2\right]$. Substituting, we find

$M\left({}_{20}^{48}\text{Ca}\right) = 28\cdot(1.008665\,\text{u}) + 20\cdot(1.007825\,\text{u}) - \left[0.43102\,\text{u}\cdot c^2/c^2\right] = 47.9681\,\text{u}$, or

$$47.9681\,\text{u}\cdot\left(\frac{931.5\,\text{MeV}}{c^2\cdot\text{u}}\right) = 4.468\times10^4\ \text{MeV}/c^2.$$

The atomic mass given in Appendix 8 is 47.952534 u. Our calculation based on a semi-empirical formula is different by about 0.03%.

27. $\lambda = \dfrac{\ln 2}{t_{1/2}} = \dfrac{\ln 2}{(5.271\ \text{y})(3.156\times10^7\ \text{s/y})} = 4.167\times10^{-9}\ \text{s}^{-1}$

$$N = \frac{R}{\lambda} = \frac{4.4\times10^7\ \text{s}^{-1}}{4.167\times10^{-9}\ \text{s}^{-1}} = 1.06\times10^{16};$$

$$m = \left(1.06\times10^{16}\right)\frac{1\ \text{mol}}{6.022\times10^{23}}\left(\frac{60\ \text{g}}{\text{mol}}\right) = 1.05\ \mu\text{g}$$

29. In general (using the definition of the mean value of a function)

$$\tau = \frac{\int_0^\infty tR(t)dt}{\int_0^\infty R(t)dt} = \frac{1}{N_0}\int_0^\infty tR(t)dt$$

because all nuclei must decay between $t = 0$ and $t = \infty$. Using $R = R_0 e^{-\lambda t}$ we have

$$\tau = \frac{R_0}{N_0}\int_0^\infty te^{-\lambda t}dt = \frac{R_0}{N_0}\%\frac{1}{\lambda^2} = \frac{\lambda N_0}{\lambda^2 N_0} = \frac{1}{\lambda} = \frac{t_{1/2}}{\ln 2}.$$

32. $\lambda = \dfrac{\ln 2}{t_{1/2}} = \dfrac{\ln 2}{(109.8\ \text{min})(60\ \text{s/min})} = 1.052\times10^{-4}\ \text{s}^{-1}$

$$R = R_0 e^{-\lambda t} = \left(1.2\times10^7\ \text{Bq}\right)\exp\left(-\left(1.052\times10^{-4}\ \text{s}^{-1}\right)(48)(3600\text{s})\right) = 0.153\ \text{Bq}.$$

© 2013 Cengage Learning. All Rights Reserved. May not be scanned, copied or duplicated, or posted to a publicly accessible website, in whole or in part.

41. $^{80}\text{Br} \rightarrow {}^{76}\text{As} + {}^{4}\text{He}:$

$Q = [79.918530 - 75.922394 - 4.002603]\text{u}\cdot\text{c}^2 = -6.0 \text{ MeV (not allowed)}$

$^{80}\text{Br} \rightarrow {}^{80}\text{Kr} + \beta^-: \; Q = (79.918530 - 79.916378)\text{u}\cdot c^2 = 2.0 \text{MeV (allowed)};$

$^{80}\text{Br} \rightarrow {}^{80}\text{Se} + \beta^+:$

$Q = (79.918530 - 79.916522 - 2(0.000549))\text{u}\cdot c^2 = 0.85 \text{ MeV (allowed)}$

$^{80}\text{Br} + \beta^- \rightarrow {}^{80}\text{Se}:$

$Q = (79.918530 - 79.916522)\text{u}\cdot c^2 = 1.9 \text{MeV (allowed)}$

42. $^{227}\text{Ac} \rightarrow {}^{223}\text{Fr} + {}^{4}\text{He}:$

$Q = [227.027747 - 223.019731 - 4.002603]\text{u}\cdot c^2 = 5.0 \text{ MeV (allowed)}$

$^{227}\text{Ac} \rightarrow {}^{227}\text{Th} + \beta^-:$

$Q = (227.027747 - 227.027699)\text{u}\cdot c^2 = 0.045 \text{ MeV (allowed, barely)}$

$^{227}\text{Ac} \rightarrow {}^{227}\text{Ra} + \beta^+:$

$Q = (227.027747 - 227.029171 - 2(0.000549))\text{u}\cdot c^2 = -2.3 \text{ MeV (not allowed)}$

$^{227}\text{Ac} + \beta^- \rightarrow {}^{227}\text{Ra}:$

$Q = (227.027747 - 227.029171)\text{u}\cdot c^2 = -1.3 \text{MeV(not allowed)}$

52. $R' = \dfrac{N(^{206}\text{Pb})}{N(^{238}\text{U})} = e^{\lambda t} - 1 = e^{\ln 2} - 1 = 1$ where the substitution for λ occurs since the time given in the problem almost exactly matches the half-life of U-238. Thus $\lambda t = (\ln(2)/t_{1/2})t = \ln(2)$. A more exact answer would be $\lambda t = (\ln(2)/4.47\times10^9)4.6\times10^9 = 1.03[\ln(2)]$ and thus the ratio would be

$R' = \dfrac{N(^{206}\text{Pb})}{N(^{238}\text{U})} = e^{\lambda t} - 1 = e^{(1.03)\ln 2} - 1 = 1.04$ revealing a slightly higher amount of lead.

53. From the A values it is clear that there are $28/4 = 7$ alpha decays. Seven alpha decays reduces Z from 92 to 78, so there must be four β^- decays in order to bring Z up to 82. There are other possible combinations of beta decays (including β^+ and electron capture), but the net result must be a change of four charge units. We would have to look at a table of nuclides to determine the exact chain(s).

56. (a) $t_{1/2} = 7.7\times10^{24}\text{y} = 2.4\times10^{32}\text{ s};$

$$R = \lambda N = \frac{\ln 2}{t_{1/2}}\frac{N_A}{0.128 \text{ kg}} = \frac{\ln 2}{2.4\times10^{32}\text{ s}}\frac{6.022\times10^{23}}{0.128 \text{ kg}} = 1.36\times10^{-8}\text{s}^{-1}\cdot\text{kg}^{-1}$$

(b) $\dfrac{10 \text{ s}^{-1}}{1.36\times10^{-8}\text{ s}^{-1}\cdot\text{kg}^{-1}} = 7.36\times10^8 \text{ kg}$. This is not a realistic sample size.

© 2013 Cengage Learning. All Rights Reserved. May not be scanned, copied or duplicated, or posted to a publicly accessible website, in whole or in part.

58. For ^{36}Ar: $B=\left[18m_n+18M\left({}^{1}\text{H}\right)-M\left({}^{36}\text{Ar}\right)\right]c^2$; $B=0.329274\ \text{u}\cdot c^2=306.7$ MeV so $B/A=8.52$ MeV/nucleon.

For ^{76}Se: $B=\left[42m_n+34M\left({}^{1}\text{H}\right)-M\left({}^{76}\text{Se}\right)\right]c^2$; $B=0.710766\ \text{u}\cdot c^2=662.1$ MeV so $B/A=8.71$ MeV/nucleon.

This is the expected result, and matches the asymmetric shape of the curve in Figure 12.6.

61. Adding charge dq to a solid sphere of radius r we have an energy change $dE=\dfrac{Q\,dq}{4\pi\epsilon_0 r}$ where $Q=\dfrac{4}{3}\rho\pi r^3$ is the charge already there and ρ is the charge density. Also we know $dq=\rho dV=4\pi\rho r^2\,dr$. Making those substitutions gives $dE=\dfrac{16\pi^2\rho^2r^4}{3\left(4\pi\epsilon_0\right)}dr$.

Integrating from 0 to R we find $\Delta E=\int_0^R\dfrac{16\pi^2\rho^2r^4}{3\left(4\pi\epsilon_0\right)}dr=\dfrac{16\pi^2\rho^2R^5}{15\left(4\pi\epsilon_0\right)}$. The charge density is $\rho=Q/V=3Q/4\pi R^3$, so $\Delta E=\dfrac{16\pi^2R^5}{15\left(4\pi\epsilon_0\right)}\left(\dfrac{3Q}{4\pi R^3}\right)^2=\dfrac{3Q^2}{5\left(4\pi\epsilon_0\right)R}=\dfrac{3\left(Ze\right)^2}{5\left(4\pi\epsilon_0\right)R}$.

67. (a) Examining the mass numbers it must come from four alpha decays of ^{238}U.

(b) From the table of nuclides the decays are: $^{222}\text{Rn}\rightarrow{}^{218}\text{Po}+{}^{4}\text{He}$; $^{218}\text{Po}\rightarrow{}^{214}\text{Pb}+{}^{4}\text{He}$; $^{214}\text{Pb}\rightarrow{}^{214}\text{Bi}+\beta^-$; $^{214}\text{Bi}\rightarrow{}^{214}\text{Po}+\beta^-$; $^{214}\text{Po}\rightarrow{}^{210}\text{Pb}+{}^{4}\text{He}$

(c) The only half-life longer than an hour is the first decay, with $t_{1/2}=3.8$ days. Therefore more than half the decays will occur in four days.

68. (a) Both ^{4}He and ^{16}O are "doubly magic"; that is both Z and N are magic numbers. These elements are more tightly bound than nuclei around them.

(b) Because ^{208}Pb has $Z=82$ and $N=126$ it is doubly magic. It is particularly stable and its binding energy is high. Other nuclei around it are unstable because of the large Coulomb energy.

(c) ^{40}Ca is doubly magic and is the most stable of the calcium isotopes. It is the heaviest nuclide with $Z=N$. ^{48}Ca is also stable and has $Z=20$ and $N=28$ which are both magic numbers. Six isotopes of calcium are stable with $Z=20$ and various values of N.

© 2013 Cengage Learning. All Rights Reserved. May not be scanned, copied or duplicated, or posted to a publicly accessible website, in whole or in part.

Chapter 13

3. The probability is $nt\sigma$: $nt\sigma = \frac{6.022\times10^{23}}{108\text{ g}}\frac{10.5\text{ g}}{\text{cm}^3}(0.2\text{ cm})\left(17\times10^{-24}\text{ cm}^2\right)=0.20$

8. (a) $Q=\left[M(^{16}\text{O})+M(^{2}\text{H})-M(^{4}\text{He})-M(^{14}\text{N})\right]\text{u}\cdot c^2=3.11\text{ MeV (exothermic)}$
   (b) $Q=\left[M(^{12}\text{C})+M(^{12}\text{C})-M(^{2}\text{H})-M(^{22}\text{Na})\right]\text{u}\cdot c^2=-7.95\text{ MeV (endothermic)}$
   (c) $Q=\left[M(^{23}\text{Na})+M(^{1}\text{H})-M(^{12}\text{C})-M(^{12}\text{C})\right]\text{u}\cdot c^2=-2.24\text{ MeV (endothermic)}$

11. (a) $Q=K_y+K_Y-K_x=1.1\text{ MeV}+8.4\text{ MeV}-5.5\text{ MeV}=4.0\text{ MeV}$
    (b) The Q value does not change for a particular reaction.

16. (a) $Q=\left[M(^{16}\text{O})+M(^{4}\text{He})-M(^{1}\text{H})-M(^{19}\text{F})\right]\text{u}\cdot c^2=-8.11\text{ MeV}$
    $K_{\text{th}}=\frac{20}{16}(8.11\text{ MeV})=10.14\text{ MeV}$
    (b) $Q=\left[M(^{12}\text{C})+M(^{2}\text{H})-M(^{3}\text{He})-M(^{11}\text{B})\right]\text{u}\cdot c^2=-10.46\text{ MeV}$
    $K_{\text{th}}=\frac{14}{12}(10.46\text{ MeV})=12.21\text{ MeV}$

18. The equation for K_{cm} is correct because we know from classical mechanics that the system is equivalent to a mass M_x+M_X moving with a speed v_{cm}, so the center of mass kinetic energy is $K_{\text{cm}}=\frac{1}{2}(M_x+M_X)v_{\text{cm}}^2$. Letting $M=M_x+M_X$ we have by conservation of momentum $Mv_{\text{cm}}=M_xv_x$, or $v_{\text{cm}}=M_xv_x/M$. Therefore

$$K_{\text{cm}}=\frac{1}{2}Mv_{\text{cm}}^2=\frac{1}{2}M\frac{M_x^2}{M^2}v_x^2=\frac{1}{2}\frac{M_x^2}{M}v_x^2$$

$$K'_{\text{cm}}=K_{\text{lab}}-K_{\text{cm}}=\frac{M_xv_x^2}{2}-\frac{1}{2}\frac{M_x^2}{M}v_x^2=\frac{M_xv_x^2}{2}\left(1-\frac{M_x}{M}\right)=\frac{M_xv_x^2}{2}\left(\frac{M-M_x}{M}\right)$$

$$K'_{\text{cm}}=K_{\text{lab}}\left(\frac{M_X}{M_x+M_X}\right)$$

19. $K'_{\text{cm}}=\frac{M_X}{M_x+M_X}K_{\text{lab}}=\frac{14}{18}(6.7\text{ MeV})=5.21\text{ MeV}$
    $E^*=\left[M(^{14}\text{N})+M(^{4}\text{He})-M(^{18}\text{F})\right]\text{u}\cdot c^2=4.41\text{ MeV}$
    $E_x=E^*+K'_{\text{cm}}=9.62\text{ MeV}$

© 2013 Cengage Learning. All Rights Reserved. May not be scanned, copied or duplicated, or posted to a publicly accessible website, in whole or in part.

22. From Chapter 12 Problem 29 we see that the mean lifetime is

$$\tau = \frac{t_{1/2}}{\ln 2} = \frac{109 \text{ ms}}{\ln 2} = 157 \text{ ms}$$

Then from Equation (13.13) $\Gamma = \frac{\hbar}{2\tau} = \frac{6.582\times10^{-16} \text{ eV}\cdot\text{s}}{2(0.157 \text{ s})} = 2.10\times10^{-15} \text{ eV}.$

^{17}Ne can decay by positron decay or electron capture.

27. (a) $m = (5.5\times10^{-4})(10^{6} \text{ kg}) = 550 \text{ kg}$

(b) $(550 \text{ kg})\frac{6.022\times10^{23}}{0.238 \text{ kg}} = 1.4\times10^{27}$ atoms

(c) $R = (6.7 \text{ kg}^{-1}\cdot\text{s}^{-1})(550 \text{ kg}) = 3700 \text{ Bq}$

(d) $(3700 \text{ s}^{-1})\frac{86400 \text{ s}}{d} = 3.2\times10^{8} \text{ d}^{-1}$

28. $$\frac{N(^{235}\text{U})}{N(^{238}\text{U})} = \frac{7.2}{993} = \frac{N_0(^{235}\text{U})e^{-\lambda_1 t}}{N_0(^{238}\text{U})e^{-\lambda_2 t}}$$

where the subscripts 1 and 2 refer to the 235 and 238 isotopes, respectively.

$$\frac{N_0(^{235}\text{U})}{N_0(^{238}\text{U})} = \frac{7.2}{993}\exp((\lambda_1 - \lambda_2)t)$$

$$(\lambda_1 - \lambda_2)t = \ln 2\left(\frac{1}{t_{1/2}(235)} - \frac{1}{t_{1/2}(238)}\right)t$$

$$= \ln 2\left(\frac{1}{7.04\times10^{8}\text{y}} - \frac{1}{4.47\times10^{9}\text{ y}}\right)(2.0\times10^{9}\text{y}) = 1.659$$

Then $\frac{N_0(^{235}\text{U})}{N_0(^{238}\text{U})} = \frac{7.2}{993}\exp((\lambda_1 - \lambda_2)t) = \frac{7.2}{993}\exp(1.659) = 0.0381$

which is more than five times higher than today. Note that with the ratio of abundances equal to 0.0381, the percentage abundances needed to give that ratio are about 3.7% and 96.3%. Natural fission reactors cannot operate because of the relatively low abundance of ^{235}U today. The exception is a type of reactor known as the Canadian Deuterium Reactor (CANDU) which uses heavy water rather than regular or light water as the moderator.

31. For uranium we assume as in the preceding problem that each fission produces 200 MeV of energy.

$$(1.0 \text{ kg})\left(\frac{6.022\times10^{23} \text{ atoms}}{0.235 \text{ kg}}\right)\frac{200 \text{ MeV}}{\text{atom}} = 5.13\times10^{26} \text{ MeV}$$

© 2013 Cengage Learning. All Rights Reserved. May not be scanned, copied or duplicated, or posted to a publicly accessible website, in whole or in part.

Converting to kWh, we find 2.30×10^7 kWh. For 1.0 kg of coal, Table 13.1 gives energy output 3×10^7 J, which is equivalent to 8.33 kWh. Therefore we see that fission produces over one million times more energy per kilogram of fuel.

34. (a) $\frac{3}{2}kT = \frac{3}{2}\left(8.617\times10^{-5}\text{ eV/K}\right)\left(300\text{ K}\right) = 3.88\times10^{-2}\text{ eV}$

(b) $\frac{3}{2}kT = \frac{3}{2}\left(8.617\times10^{-5}\text{ eV/K}\right)\left(15\times10^{6}\text{ K}\right) = 1.94\text{ keV}$

39. (a) This problem is similar to Example 13.9. First we will determine the Coulomb potential energy that must be overcome, using Equation (12.2) to determine the radius.

$r = r_0 A^{1/3} = 1.2\times10^{-15}\text{ m}\left(12\right)^{1/3} = 2.7\times10^{-15}\text{ m}$

Therefore the Coulomb barrier is:

$$V = \frac{q_1 q_2}{4\pi\epsilon_0 r} = \frac{\left(9\times10^{9}\text{N}\cdot\text{m}^2/\text{C}^2\right)\left(6\right)^2\left(1.6\times10^{-19}\text{C}\right)^2}{2.7\times10^{-15}\text{m}} = 3.1\times10^{-12}\text{J}.$$

We need at least this much kinetic energy to overcome the Coulomb barrier. We set this value equal to the thermal energy, namely $\frac{3}{2}kT$.

$$T = \frac{2V}{3k} = \frac{2\left(3.1\times10^{-12}\text{J}\right)}{3\left(1.38\times10^{-23}\text{J/K}\right)} = 1.5\times10^{11}\text{K}$$

(b) We use Equation (13.7) to find the Q value for the reaction:

$$Q = \left[2M(^{12}\text{C}) - M(^{24}\text{Mg})\right]\text{u}\cdot c^2 = \left[2(12.0) - 23.985042\right]\text{u}\cdot c^2 = 13.9\text{MeV}$$

This includes the energy of the γ ray in the released energy.

47. (a) With a half-life of 6.01 h, not much of the original material remains after one week.

(b) $R = R_0 e^{-\lambda t} = \left(10^{11}\text{ Bq}\right)\exp\left(-\frac{\ln 2}{6\text{h}}(216\text{ h})\right) = 1.46\text{ Bq}$

(c) $R = R_0 e^{-\lambda t} = \left(0.9\times10^{11}\text{ Bq}\right)\exp\left(-\frac{\ln 2}{6\text{h}}(96\text{ h})\right) = 1.37\times10^{6}\text{Bq}$

58. (a) $\left(1\text{ kg}\right)\frac{6.022\times10^{23}}{\text{mol}}\frac{1\text{mol}}{0.001\text{ kg}}\frac{13.6\text{ eV}}{1\text{ atom}} = 8.19\times10^{27}\text{ eV}$

© 2013 Cengage Learning. All Rights Reserved. May not be scanned, copied or duplicated, or posted to a publicly accessible website, in whole or in part.

(b) $(1\text{ kg})\dfrac{6.022\times10^{23}}{\text{mol}}\dfrac{1\text{ mol}}{0.002\text{ kg}}\dfrac{2.224\times10^{6}\text{ eV}}{1\text{ atom}}=6.70\times10^{32}\text{ eV}$

(c) $(1\text{ kg})\dfrac{1\text{ u}}{1.66\times10^{-27}\text{ kg}}\dfrac{931.49\times10^{6}\text{MeV}}{\text{u}}=5.61\times10^{35}\text{ eV}$

(d) The results are as expected. The annihilation process in part (c) is the complete conversion of matter into energy, which maximizes energy yield. The energy in (b) is much greater than the energy in (a), because nuclear binding energies are so much larger than atomic binding energies.

59. (a) $Q=K_{\text{out}}-K_{\text{in}}=86.63\text{ MeV }-100\text{ MeV }=-13.37\text{ MeV}$

This agrees very nearly with the Q computed using atomic masses (assuming all the masses are known).

$Q=\left[M(^{18}\text{O})+M(^{30}\text{Si})-M(^{14}\text{O})-M(^{34}\text{Si})\right]\text{u}\cdot c^2=-13.27\text{ MeV}$

(b) $M=-\dfrac{Q}{c^2}+M(^{18}\text{O})+M(^{30}\text{Si})-M(^{14}\text{O})=33.979\text{ u}$

which agrees very nearly with the value for the mass of ^{34}Si in Appendix 8.

63. $m(^{40}\text{K})=(65\text{ kg})(0.0035)(0.00012)=2.73\times10^{-5}\text{ kg}$

$N=\left(2.73\times10^{-5}\text{ kg}\right)\dfrac{6.022\times10^{23}}{0.040\text{ kg}}=4.11\times10^{20}$

$R=\lambda N=\left(\dfrac{\ln 2}{1.28\times10^{9}\text{ y}}\right)\left(\dfrac{1\text{ y}}{3.156\times10^{7}\text{ s}}\right)\left(4.11\times10^{20}\right)=7050\text{ Bq}$

The beta activity is $(7050\text{ Bq})(0.893)=6300\text{ Bq}$.

© 2013 Cengage Learning. All Rights Reserved. May not be scanned, copied or duplicated, or posted to a publicly accessible website, in whole or in part.

Chapter 14

1. By conservation of momentum the photons must have the same energy. For each one

$$E = hf = mc^2 \text{ and } f = \frac{mc^2}{h} = \frac{938.27\times10^6 \text{ eV}}{4.136\times10^{-15} \text{ eV}\cdot\text{s}} = 2.27\times10^{23} \text{ Hz.}$$

2. As in the text (see Example 14.1), let $mc^2 = \hbar c / R$, so

$$R = \frac{\hbar c}{mc^2} = \frac{197.3 \text{ eV}\cdot\text{nm}}{140\times10^6 \text{ eV}} = 1.41\times10^{-6} \text{ nm} = 1.41 \text{ fm.}$$

4. We know that $\lambda \le D$ and the problem says specifically to choose $\lambda = 0.10D$ with the diameter $D = 0.15$ fm. Therefore $\lambda = 0.1(1.5 \text{ fm}) = 0.15$ fm. We know from de Broglie's relationship that $p = h/\lambda$ and we can determine the kinetic energy from this momentum as follows: $E^2 = \left(K + mc^2\right)^2 = (pc)^2 + \left(mc^2\right)^2$

$$K = \sqrt{(pc)^2 + \left(mc^2\right)^2} - mc^2 = \sqrt{\left(\frac{hc}{0.15\text{fm}}\right)^2 + \left(mc^2\right)^2} - mc^2$$

Electron:

$$K = \sqrt{\left(\frac{1239.8 \text{ eV}\cdot\text{nm}}{0.15 \text{ fm}\left(10^{-6}\text{nm/1 fm}\right)}\right)^2 + \left(0.511\times10^6\text{eV}\right)^2} - 0.511\times10^6\text{eV} = 8.26 \text{ GeV}$$

Proton:

$$K = \sqrt{\left(\frac{1239.8 \text{ eV}\cdot\text{nm}}{0.15 \text{ fm}\left(10^{-6}\text{nm/1 fm}\right)}\right)^2 + \left(938.27\times10^6\text{eV}\right)^2} - 938.27\times10^6\text{eV} = 7.38 \text{ GeV}$$

7. We use Equation (14.4) to estimate the range of the force.

$$R = \frac{\hbar c}{2mc^2} = \frac{197.33 \text{ eV}\cdot\text{nm}}{2\left(150\times10^9\text{eV}\right)} = 6.58\times10^{-10}\text{nm} = 6.58\times10^{-19}\text{m.}$$

15. Let subscript 1 refer to the Σ, subscript 2 to the Λ, and no subscript to the photon. From conservation of momentum $|p| = |p_2| = E/c$. From conservation of energy

$$m_1c^2 = \sqrt{p_2^2c^2 + \left(m_2c^2\right)^2} + E = \sqrt{E^2 + \left(m_2c^2\right)^2} + E \text{, so}$$

$$E = \frac{\left(m_1c^2\right)^2 - \left(m_2c^2\right)^2}{2m_1c^2} = \frac{(1193 \text{ MeV})^2 - (1116 \text{ MeV})^2}{2(1193 \text{ MeV})} = 74.5 \text{ MeV.}$$

© 2013 Cengage Learning. All Rights Reserved. May not be scanned, copied or duplicated, or posted to a publicly accessible website, in whole or in part.

18. $n\ (udd)$: $q=\frac{2e}{3}-\frac{e}{3}-\frac{e}{3}=0;$ $B=3\left(\frac{1}{3}\right)=1;$ $S=3(0)=0$

$\Sigma^+\ (uus)$: $q=\frac{2e}{3}+\frac{2e}{3}-\frac{e}{3}=e;$ $B=3\left(\frac{1}{3}\right)=1;$ $S=0+0-1=-1$

$\Lambda_C^+\ (udc)$: $q=\frac{2e}{3}-\frac{e}{3}+\frac{2e}{3}=e;$ $B=3\left(\frac{1}{3}\right)=1;$ $S=0+0+0=0$

19. $\pi^+\ (u\bar{d})$: $q=\frac{2e}{3}+\frac{e}{3}=e;$ $B=\frac{1}{3}-\frac{1}{3}=0;$ $S=0+0=0$

$K^+\ (u\bar{s})$: $q=\frac{2e}{3}+\frac{e}{3}=e;$ $B=\frac{1}{3}-\frac{1}{3}=0;$ $S=0+1=1$

$D^0\ (c\bar{u})$: $q=\frac{2e}{3}-\frac{2e}{3}=e;$ $B=\frac{1}{3}-\frac{1}{3}=0;$ $S=0+0=0$

24. We use Table 14.5 for quark properties and Table 14.4 or Table 14.6 to identify the hadrons. The spin is 1/2 for all quarks and antiquarks.

(a) $c\bar{d}$; spin is 0 or 1; charge is 1; baryon number is 0; $C = 1$; $S, B, T = 0$. This is a D^+ meson.
(b) uds; spin is 1/2 or 3/2; charge is 0; baryon number is 1; $S=-1$; $C,B,T=0$. This could be a Σ^0 or a Λ baryon.
(c) $\overline{sss}$; spin is 1/2 or 3/2; baryon number is -1; charge is 1; $S = 3$; $C,B,T=0$. This is a Ω^+ baryon.
(d) $\bar{c}d$; spin is 0 or 1; charge is -1; baryon number is 0; $C=-1$; $S,B,T=0$. This is a D^- meson.

29. We begin with Equation (14.10). Since $K \gg mc^2$, this simplifies to $E_{cm}=\sqrt{2mc^2K_{lab}}$. From the problem statement, we know that $E_{cm}=2K$ so $E_{cm}=2K=\sqrt{2mc^2K_{lab}}$. This can be rearranged to find $4K^2=2mc^2K_{lab}$ or $K_{lab}=\frac{2K^2}{mc^2}$.

34. (a) $$p=\frac{\sqrt{E^2-E_0^2}}{c}=\frac{\sqrt{(948.27\text{ MeV})^2-(938.27\text{ MeV})^2}}{c}=137.35\text{ MeV}/c$$

$$R=\frac{p}{qB}=\frac{137.35\times10^6\text{ eV}\left(1.602\times10^{-19}\frac{\text{J}}{\text{eV}}\right)}{(2.998\times10^8\text{ m/s})(e)(1.4\text{ T})}=0.327\text{ m}.$$

© 2013 Cengage Learning. All Rights Reserved. May not be scanned, copied or duplicated, or posted to a publicly accessible website, in whole or in part.

(b) From Equation (14.9) we have $f = \dfrac{eB}{2\pi m}\sqrt{1-v^2/c^2} = \dfrac{eB}{2\pi m\gamma}$ with

$$\gamma = \frac{E}{E_0} = \frac{948.27\text{ MeV}}{938.27\text{ MeV}} = 1.0107$$

Therefore

$$f = \frac{(e)(1.4\text{ T})}{2\pi\left(938.27\times10^6\text{ eV}\right)(1.0107)}\left(2.998\times10^8\text{ m/s}\right)^2 = 2.11\times10^7\text{Hz.}$$

37. As in the preceding problem $E_{\text{cm}} = \sqrt{\left(m_1c^2+m_2c^2\right)^2+2Km_2c^2} = \sqrt{\left(2mc^2\right)^2+2Km_2c^2}$.

(a) If $K \ll mc^2$ we neglect K, so $E_{\text{cm}} \approx 2mc^2$.

(b) If $K \gg mc^2$ we neglect the first term and $E_{\text{cm}} \approx \sqrt{2Km_2c^2} = \sqrt{2Kmc^2}$.

In (a) we interpret the result to mean that at very low energies there is no extra energy available (beyond the masses of the two original particles). In (b) we see that the available center of mass energy increases only in proportion to $\sqrt{K}$, thus illustrating the great advantage of colliding beam experiments over fixed target experiments.

49. (a) For a stationary target the sum of the rest energies of the products equals the total center of mass energy, so

$$E_{\text{cm}} = \sqrt{2E_0\left(2E_0+K\right)} = \left(m_p+m_\Lambda+m_K\right)c^2$$
$$= 938\text{ MeV} + 1116\text{ MeV} + 494\text{ MeV} = 2548\text{ MeV}$$

Rearranging we have $\dfrac{E_{\text{cm}}^2}{2E_0} = 2E_0 + K$;

$$K = \frac{E_{\text{cm}}^2}{2E_0} - 2E_0 = \frac{(2548\text{ MeV})^2}{2(938\text{ MeV})} - 2(938\text{ MeV}) = 1585\text{ MeV.}$$

(b) In a colliding beam experiment the total momentum is zero, and we have by conservation of energy $2E_0 + 2K = E_0 + \left(m_\Lambda + m_K\right)c^2$;

$$K = \frac{-E_0+\left(m_\Lambda+m_K\right)c^2}{2} = \frac{-938\text{ MeV}+1116\text{ MeV}+494\text{ MeV}}{2} = 336\text{ MeV.}$$

52. (a) baryon number and electron lepton number not conserved
(b) not allowed; charge is not conserved
(c) allowed
(d) allowed

© 2013 Cengage Learning. All Rights Reserved. May not be scanned, copied or duplicated, or posted to a publicly accessible website, in whole or in part.

53. (a) strangeness is not conserved
    (b) charge is not conserved
    (c) baryon number is not conserved
    (d) strangeness is not conserved

© 2013 Cengage Learning. All Rights Reserved. May not be scanned, copied or duplicated, or posted to a publicly accessible website, in whole or in part.

Chapter 15

1. From Newton's second law we have for a pendulum of length L:

$F = m_G g \sin\theta = m_I a = m_I L \dfrac{d^2\theta}{dt^2}$; $\dfrac{d^2\theta}{dt^2} = \dfrac{m_G g}{m_I L}\sin\theta \approx \dfrac{m_G g}{m_I L}\theta$, where we have made the small-angle approximation $\sin\theta \approx \theta$. This is a simple harmonic oscillator equation with solution $\theta = \theta_0 \cos(\omega t)$ where θ_0 is the amplitude and the angular frequency is $\omega = \sqrt{\dfrac{m_G g}{m_I L}}$. The period of oscillation is $T = \dfrac{2\pi}{\omega} = 2\pi\sqrt{\dfrac{m_I L}{m_G g}}$. Therefore two masses with different ratios m_I / m_G will have different small-amplitude periods.

3. Beginning with Equation (15.4)

$$\frac{\Delta f}{f} = -\frac{GM}{c^2}\left(\frac{1}{r_1} - \frac{1}{r_2}\right) = -\frac{GM}{r_1 r_2 c^2}(r_2 - r_1) = \frac{GM}{r_1 r_2 c^2}(r_1 - r_2).$$

Use $r_1 - r_2 = H$ and let $r_1 \approx r_2 = r$. From classical mechanics $g = GM / r^2$, so $\dfrac{\Delta f}{f} = \dfrac{gH}{c^2}$.

6. Using the mass and the radius of the neutron star and given that the detection is at a very large distance,

$$\frac{\Delta f}{f} = -\frac{GM}{c^2}\left(\frac{1}{r_1} - \frac{1}{r_2}\right) = \frac{GM}{rc^2} = \frac{\left(6.673\times10^{-11}\,\text{m}^3\cdot\text{kg}^{-1}\cdot\text{s}^{-2}\right)\left(5.8\times10^{30}\,\text{kg}\right)}{\left(1.0\times10^4\ \text{m}\right)\left(2.998\times10^8\,\text{m/s}\right)^2} = 4.31\times10^{-1}$$

The wavelength is affected by the same factor, so the redshift at the given wavelength is $\Delta\lambda = \left(4.31\times10^{-1}\right)\left(550\ \text{nm}\right) = 240$ nm, where we have rounded to the two significant figures given in the problem. Thus the new wavelength is 550 nm + 240 nm = 790 nm, which is above the visible range in the infrared.

7. Using the mass and radius of the sun

$$\frac{\Delta f}{f} = \frac{GM}{rc^2} = \frac{\left(6.673\times10^{-11}\ \text{m}^3\cdot\text{kg}^{-1}\cdot\text{s}^{-2}\right)\left(1.99\times10^{30}\ \text{kg}\right)}{\left(6.96\times10^8\ \text{m}\right)\left(2.998\times10^8\ \text{m/s}\right)^2} = 2.123\times10^{-6}.$$

The redshift of the wavelength is affected by the same factor, so the redshift at the two wavelengths equals: $\Delta\lambda = \left(2.123\times10^{-6}\right)\left(400\ \text{nm}\right) = 8.49\times10^{-4}$ nm and $\Delta\lambda = \left(2.123\times10^{-6}\right)\left(700\ \text{nm}\right) = 1.49\times10^{-3}$ nm.

© 2013 Cengage Learning. All Rights Reserved. May not be scanned, copied or duplicated, or posted to a publicly accessible website, in whole or in part.

13. $r_s = \dfrac{2GM}{c^2} = \dfrac{2\left(6.673\times10^{-11}\ \text{m}^3\cdot\text{kg}^{-1}\cdot\text{s}^{-2}\right)\left(1.90\times10^{27}\ \text{kg}\right)}{\left(2.998\times10^8\ \text{m/s}\right)^2} = 2.82\ \text{m}$

14. (a) From Equation (15.7) we have

$$T = \frac{\hbar c^3}{8\pi kGM} = \frac{\left(1.0546\times10^{-34}\ \text{J}\cdot\text{s}\right)\left(2.998\times10^8\ \text{m/s}\right)^3}{8\pi\left(1.381\times10^{-23}\ \text{J/K}\right)\left(6.673\times10^{-11}\ \text{m}^3\cdot\text{kg}^{-1}\cdot\text{s}^{-2}\right)\left(1.99\times10^{30}\ \text{kg}\right)}$$
$$= 6.17\times10^{-8}\,\text{K}$$

(b) Using Equation (15.7)

$$T = \frac{\hbar c^3}{8\pi kGM} = \frac{\left(1.05\times10^{-34}\ \text{J}\cdot\text{s}\right)\left(3.0\times10^8\ \text{m/s}\right)^3}{8\pi\left(1.38\times10^{-23}\ \text{J/K}\right)\left(6.67\times10^{-11}\ \text{N}\cdot\text{m}^2/\text{kg}^2\right)\left(6\times10^9\right)\left(2.0\times10^{30}\ \text{kg}\right)}$$
$$= 1.0\times10^{-17}\,\text{K}$$

Such low temperatures, especially for part (b) would make the object extremely difficult to observe.

17. (a) We use the results of Example 15.3 that give the time in terms of the mass. Rearranging the equation, using the age of the universe as 13.7 billion years, and using the value of α from problem 16, we have

$$M_0 = (3\alpha t)^{1/3} = \left[3\left(3.965\times10^{15}\text{kg}^3/\text{s}\right)\left(13.7\times10^9\,\text{y}\frac{3.156\times10^7\,\text{s}}{1\text{y}}\right)\right]^{1/3} = 1.73\times10^{11}\text{kg}.$$

(b) Current evidence for the smallest black holes require a mass of about 5 to 20 times the mass of the Sun. The mass from part (a) is only $M_0 \approx 10^{-19} M_{\text{Sun}}$ which is too small to form a black hole.

22. Set the change in the photon's energy equal to the change in gravitational potential energy: $\Delta E = h\Delta f = -\dfrac{GMm}{r_1} - \left(-\dfrac{GMm}{r_2}\right) = -GMm\left(\dfrac{1}{r_1} - \dfrac{1}{r_2}\right)$ where M is the mass of the earth and m is the equivalent mass of the photon. Now $m = E/c^2 = hf/c^2$, so

$$h\Delta f = -\frac{GMhf}{c^2}\left(\frac{1}{r_1} - \frac{1}{r_2}\right) \text{ and } \frac{\Delta f}{f} = -\frac{GM}{c^2}\left(\frac{1}{r_1} - \frac{1}{r_2}\right).$$

© 2013 Cengage Learning. All Rights Reserved. May not be scanned, copied or duplicated, or posted to a publicly accessible website, in whole or in part.

Chapter 16

1. $$1\text{ pc} = \frac{1\text{ au}}{\tan 1''}\left(1.496\times10^{11}\text{ m/au}\right) = 3.086\times10^{16}\text{ m}$$

   $$1\text{ ly} = \left(2.9979\times10^{8}\text{ m/s}\right)\left(365.25\text{d/y}\right)\left(86400\text{ s/d}\right) = 9.461\times10^{15}\text{ m}$$

   $$1\text{ pc} = \frac{3.086\times10^{16}\text{ m}}{9.461\times10^{15}\text{m/ly}} = 3.26\text{ ly}$$

2. As in Example 16.3 we have $7.0 = \exp\left(\Delta mc^2 / kT\right)$ so $\ln 7.0 = \Delta mc^2 / kT$. Then

   $$T = \frac{\Delta mc^2}{k\ln 7.0} = \frac{939.56563\text{ MeV} - 938.27231\text{ MeV}}{\left(8.617\times10^{-11}\text{ MeV/K}\right)\left(\ln 7.0\right)} = 7.71\times10^{9}\text{ K.}$$

7. The $\pi^+\left(E_0 = 140\text{ MeV}\right)$ is more massive than the $\pi^0\left(E_0 = 135\text{ MeV}\right)$, so the π^+ would have been formed for a longer time. With $\Delta mc^2 = k\Delta T$ we have

   $$\Delta T = \frac{\Delta mc^2}{k} = \frac{5\text{ MeV}}{8.617\times10^{-11}\text{ MeV/K}} = 5.80\times10^{10}\text{ K.}$$

8. Set the deuteron binding energy 2.22 MeV equal to kT:

   $$T = \frac{2.22\text{ MeV}}{k} = \frac{2.22\text{ MeV}}{8.617\times10^{-11}\text{MeV/K}} = 2.58\times10^{10}\text{ K.}$$

14. We begin with Equation (16.1):

    $$v = HR = \frac{71\text{ km/s}}{\text{Mpc}}\left(4.0\times10^{9}\text{ly}\right)\left(\frac{1\text{ Mpc}}{3.26\times10^{6}\text{ly}}\right) = 8.71\times10^{4}\text{km/s} = 0.29c.$$

15. $$R = \frac{v}{H} = \frac{15000\text{ km/s}}{71\text{ km/s/Mpc}} = 211\text{ Mpc or about } 689\text{ Mly}.$$

19.

    (a) From Equation (16.18), for a redshift of 8.6 we have $\Delta\lambda / \lambda_0 = 8.6$, so

    $8.6 = \sqrt{\frac{1+\beta}{1-\beta}} - 1$ which can be solved to find $\beta = 0.979$ or $v = 0.979c$.

    (b) $$R = \frac{v}{H} = \frac{0.979\left(299790\text{ km/s}\right)}{71\text{ km/s/Mpc}} = 4130\text{ Mpc};$$

    $$4130\text{ Mpc}\left(\frac{3.26\text{ Mly}}{1\text{ Mpc}}\right) = 13.5\text{ Gly.}$$

© 2013 Cengage Learning. All Rights Reserved. May not be scanned, copied or duplicated, or posted to a publicly accessible website, in whole or in part.

25. For $H = 64\ \text{km/s/Mpc}$ we have

$$H = (64000\ \text{m/s/Mpc})\left(\frac{1\ \text{Mpc}}{3.086\times10^{22}\ \text{m}}\right) = 2.07\times10^{-18}\ \text{s}^{-1}$$

$$\rho_c = \frac{3H^2}{8\pi G} = \frac{3\left(2.07\times10^{-18}\ \text{s}^{-1}\right)^2}{8\pi\left(6.67\times10^{-11}\ \text{m}^3\cdot\text{kg}^{-1}\cdot\text{s}^{-2}\right)} = 7.7\times10^{-27}\ \text{kg/m}^3.$$

For $H = 78\ \text{km/s/Mpc}$ we have

$$H = (78000\ \text{m/s/Mpc})\left(\frac{1\ \text{Mpc}}{3.086\times10^{22}\ \text{m}}\right) = 2.53\times10^{-18}\ \text{s}^{-1}$$

$$\rho_c = \frac{3H^2}{8\pi G} = \frac{3\left(2.53\times10^{-18}\ \text{s}^{-1}\right)^2}{8\pi\left(6.67\times10^{-11}\ \text{m}^3\cdot\text{kg}^{-1}\cdot\text{s}^{-2}\right)} = 1.2\times10^{-26}\ \text{kg/m}^3.$$

35. Redshift $= \dfrac{\Delta\lambda}{\lambda_0} = \dfrac{580.0\ \text{nm} - 121.6\ \text{nm}}{121.6\ \text{nm}} = 3.77 = \sqrt{\dfrac{1+\beta}{1-\beta}} - 1$

Therefore $\beta = 0.916$ and $v = 0.916c.$

36. By definition $\dfrac{\Delta\lambda}{\lambda_0} = z$. From Equation (16.18) we know $\dfrac{\Delta\lambda}{\lambda_0} = \sqrt{\dfrac{1+\beta}{1-\beta}} - 1$;

this is equivalent to $\dfrac{\Delta\lambda}{\lambda_0} + 1 = z + 1 = \sqrt{\dfrac{1+\beta}{1-\beta}}$.

40. We begin with Equation (16.21):

$$\rho_c = \frac{3H^2}{8\pi G} = \left(\frac{H\ \text{in (km/s)/Mpc}}{100}\right)^2 \frac{3\times10^4}{8\pi\left(6.6726\times10^{-11}\ \text{m}^3\cdot\text{kg}^{-1}\cdot\text{s}^{-2}\right)}$$

$$= \left(\frac{H}{100}\right)^2 \left(1.789\times10^{13}\text{km}^2\cdot\text{s}^{-2}\cdot\text{Mpc}^{-2}\cdot\text{m}^{-3}\cdot\text{kg}^{-1}\cdot\text{s}^2\right)\left(\frac{\text{Mpc}}{3.086\times10^{22}\text{m}}\right)^2\left(10^3\frac{\text{m}}{\text{km}}\right)^2$$

$$= \left(\frac{H}{100}\right)^2 \left(1.88\times10^{-26}\text{kg/m}^3\right) = \left(\frac{H}{100}\right)^2 \left(1.88\times10^{-29}\text{g/m}^3\right)$$

© 2013 Cengage Learning. All Rights Reserved. May not be scanned, copied or duplicated, or posted to a publicly accessible website, in whole or in part.